The Proactionary Imperative

The Proactionary Imperative

A Foundation for Transhumanism

Steve Fuller
Department of Sociology, University of Warwick, UK

and

Veronika Lipińska
Department of Law, University of Warwick, UK

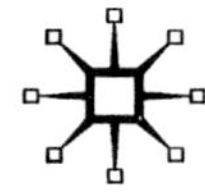

First published 2014 by
PALGRAVE MACMILLAN

Palgrave Macmillan in the UK is an imprint of Macmillan Publishers Limited, registered in England, company number 785998, of Houndmills, Basingstoke, Hampshire RG21 6XS.

Palgrave Macmillan in the US is a division of St Martin's Press LLC, 175 Fifth Avenue, New York, NY 10010.

Palgrave Macmillan is the global academic imprint of the above companies and has companies and representatives throughout the world.

Palgrave® and Macmillan® are registered trademarks in the United States, the United Kingdom, Europe and other countries.

ISBN 978–1–137–30297–7 hardback
ISBN 978–1–137–43309–1 paperback

This book is printed on paper suitable for recycling and made from fully managed and sustained forest sources. Logging, pulping and manufacturing processes are expected to conform to the environmental regulations of the country of origin.

A catalogue record for this book is available from the British Library.

A catalog record for this book is available from the Library of Congress.

Typeset by MPS Limited, Chennai, India.

Steve Fuller dedicates this book to Veronika Lipińska, whose interest, insight and persistence made possible the development of a few ideas into a full-blown ideology

Veronika Lipińska dedicates this book to her mother, Zofia Lipinska, her grandmother, Janina Chlopczynska and to her best friend, Paulina Torniewska

Contents

Tables

Introduction

The Proactionary Imperative aims to provide a comprehensive intellectual basis for the emerging progressive movement of *transhumanism*. We understand 'transhumanism' quite specifically as the indefinite promotion of the qualities that have historically distinguished humans from other creatures, which amount to our seemingly endless capacity for self-transcendence, our 'god-like' character, if you will. Whereas self-declared transhumanists tend to stress the libertarian side of this aspiration (i.e. the freedom to be whoever one wishes, aka 'morphological freedom'), we shift the emphasis to the collective normative implications of this freedom: What does it mean to act responsibly in a world where we are aiming to increase our power along many dimensions at once? An adequate answer to this question demands more than a personal ethic; it requires a political ideology, one that draws on the resources of *both* science and theology – both genetics and Genesis – as we take seriously yet open-mindedly the proposition that we are touched by the divine.

This is not to deny that we are the products of natural evolutionary forces but to circumscribe the significance of that fact. Just as democracy has transcended its origins in classical Athens to embrace a scale and scope – and diversity of forms – that would have been inconceivable to the ancients, something similar may be happening now to 'the human', the name of a creature who began life as an exotic upright ape but need not continue in that form in order to realize its full potential. Thus, we take seriously – albeit peripherally in this book – Ray Kurzweil's quest for the 'singularity', whereby accelerating computational power turns humanity into the vehicle of a cosmic intellectual revolution. Transhumanists rightly draw attention to the rapid improvements in our knowledge of the human genome and the equally rapid expansion in our digital technological capacity. These constitute the new capital base that will serve as the platform for enhancing, if not outright replacing, aspects of our evolutionary heritage. However, largely to streamline what will strike many readers as a counter-intuitive discussion, we shall presume until the final section – 'The Proactionary Manifesto' – that human biology provides the material platform for transhumanist projects in the foreseeable future.

The difference between *enhancing* and *replacing* – that is, living a thousand years as some sort of upright ape *versus* humanity's exemplary qualities surviving indefinitely in a medium indifferent to our simian origins – is a political debate that may not begin in earnest for another century. Nevertheless, both sides of this debate begin with a secular version of the idea that *Homo sapiens* is in need of 'absolution', the Christian word for the removal of sin. The word literally refers to water's capacity to remove stains from cloth. The 'stains' in this case are features of our socio-biological past that may have been necessary to get us to where we are but hold us back from our future promise. It is in this sense that Hegel's philosophy of history is rightly seen as a kind of 'absolute idealism'. Proactionaries offer a more materialist take on this vision but one that deviates significantly from that of Hegel's most dutiful materialist follower, Karl Marx. In particular, as will become clear in Chapter 3, we revisit the still taboo topic of *eugenics* in its original conception, namely, as the foundational science of human capital.

An honest transhumanist appraisal of the dominant theory of evolution, Neo-Darwinism, is that it portrays our species as too 'path-dependent', in the economists' phrase. In other words, we succeed by crowding out other species and then become so accustomed to the world we have created that we are easily eliminated once our dominion is disrupted by factors beyond our control, quite possibly as a long-term unintended consequence of our own actions. To be sure, the alarmist side of the Green movement is already deploying this narrative to explain our ultimate demise from global warming, though the more historically nuanced among the alarmists project a future in which China is the ultimate beneficiary (Oreskes and Conway 2013). Nevertheless, it is impossible to be a transhumanist and accept Neo-Darwinism's game plan, which ends with the extinction of our species and our matter reabsorbed into an almighty Nature that then recycles it, karma-like, for the benefit of future beings (cf. Fuller 2006: chap. 11).

In this context, we can distinguish two diametrically opposed responses to the shortfalls of the modern humanist world-view: *posthumanism* and *transhumanism* (Fuller 2012: chap. 3). On the one hand, posthumanists are inclined towards 'humbling' human ambitions in the face of nature's manifestly diverse and precarious character, which sometimes veers into outright misanthropy. Posthumanists see transhumanists as engaged in an especially dangerous version of 'the denial of death' that has dogged the Western imagination since Nietzsche proclaimed the death of the Abrahamic deity that had promised immortal life to humans (Becker 1973). We do not wish to minimize the seriousness of this charge – and not only because the originator of the 'proactionary principle', the transhumanist philosopher, Max More, is now CEO of the main cryonics firm, Alcor. There is also the historic association between the quest for immortality and the justification for secular evil. However, answering that charge must await another work. On the other hand, we agree with transhumanists in diagnosing humanism's failure in terms of insufficient follow-through on its own quite reasonable ambitions. This sharp contrast in post- and trans- humanist attitudes explains the repeated focus of this book on attitudes towards *risk*. *The Proactionary Imperative* is about embracing risk as constitutive of what it means to be human: *Better to give hostage to fortune than be captive to the past*. This was the original context in which Max More (2005) coined the 'proactionary principle' – as a foil to the more widely known 'precautionary principle' that would have us minimize risk in the name of global survival, a signature posthumanist stance.

Proactionaries are not *primarily* interested in ensuring that every kind of being currently on the planet survives or enjoys the same standard of existence, a state that postmodern philosophers often portray as a levelled ontological playing field (e.g. Latour 2004). To the proactionary, there is nothing intrinsically valuable in this sense of 'equality', despite its reputation as a posthumanist utopia. On the contrary, it looks like the enforcement of what the former Wall Street trader and self-styled 'risk engineer' Nicholas Taleb (2012) would call a 'fragile' approach to the ecology that fails to recognize the creative power of destruction in both natural and human history. But that does not mean that we throw all caution to the wind. The classic welfare state concern – nowadays increasingly extended to environment – about 'quality of life' may function as a *secondary* constraint on the pursuit of what really matters to the proactionary, namely, the full realization of human potential. However, this is something to which each individual human and non-human may contribute rather differently. (In the case of 'non-humans', we have in mind strategies of species preservation – typically DNA archiving – for purposes of scientific research, bioprospecting, biomimetic technologies and 'natural capitalism' more generally, as well as 'species uplift', namely, the enhancement of animal capacities in a more anthropic direction.) Here we concur with the default libertarianism of most transhumanists, who encourage risky personal experimentation. However, in addition, we insist on public access to the results of such experiments and advocate legal arrangements that would encourage people to invest themselves or their capital in risky scientific experiments.

Secular readers might wonder why we continue to fight against Darwin. It is because Charles Darwin was not only a distinguished contributor to science but also a talisman for anyone who lost their faith through science. This take on Darwin, originally diagnosed by the great Presbyterian theologian Benjamin Warfield (1888) as 'affective decline', continues to resonate strongly on both sides of the evolution controversy. It effectively creates a 'good fences make good neighbours' policy – what Stephen Jay Gould (1999) dubbed 'non-overlapping magisteria' – that prevents honest scientific and theological intercourse. This counsel of mutual respect constitutes a pernicious etiquette that inhibits scientists from declaring any deep sense of purpose that might inform their research, which to the naked eye easily appears esoteric, inconclusive if not downright risky. Indeed, in lieu of declarations of purpose, scientists are prone to generate blather about 'curiosity', a term of piety for secular thinkers that just about succeeds in licensing free inquiry, albeit by failing to distinguish the pursuit of science from that of gossip (cf. Fuller 2008a: chap. 2).

In contrast, our own view was captured by the founder of cybernetics, Norbert Wiener, who declared, 'Science is a way of life which can only flourish when men [*sic*] are free to have faith'

(Wiener 1967: 263–4). His 'faith' had less to do with organized religion than with the ultimate intelligibility of reality, which in turn justifies the effort and significance attached to scientific inquiry. This attitude reflects the Unitarian orientation of the pioneers of computational adventurism – not only Wiener but also, before him, George Boole (of 'Boolean algebra' fame) and, after him, artificial intelligence pioneer Herbert Simon and prosthetic technology entrepreneur Ray Kurzweil. By 'Unitarian' we mean the idea that each person's connection to the original creative deity is direct and personal – which is to say, not requiring any clerical mediation for its realization. From the Unitarian standpoint, the established Christian idea that God consists in 'three persons' (Father, Son and Holy Spirit) is needlessly bureaucratic, as Unitarians believe that we 'always already' have God within us but perhaps not the means to realize our divine potential. However, the requisite means are to be found not in religious services but scientific inquiry.

The significance of this theology as a deep tradition of dissent within the history of Christianity is pursued in Chapter 2, but in Wiener's hands, this tradition was given a Cold War spin as he held that nature does not engage in the sorts of tricks designed to thwart our inquiries that fellow humans sometimes do. This was a formative insight in the origin of 'social epistemology' (Fuller 1988: chap. 2). While endorsing Wiener's general line of thought, we admit one sinister implication is that the secret to fathoming humans may lie in treating them as 'natural' in some epistemologically luminous sense in which systematic attention to spontaneously generated behaviours disarms their potential for trickery (Mirowski 2002: 54–68). It was in this context that B.F. Skinner's radical behaviourism appeared to hold so much promise to people in Wiener's generation, which in our own time is more likely to be connected with brain scanning.

If 'evolution' stood simply for an empirically observable process the ultimate causes of which are subject to legitimate dispute, it is unlikely that the science–religion controversy would have acquired its global cultural significance. However, 'Darwinism' stands for a profound trade-off between science and faith, which sociologists have described as a 'demystification' or a 'disenchantment' of the world, depending on whether one's intellectual marching orders are taken from Karl Marx or Max Weber (Proctor 1991: chap. 3). To their credit, historians and philosophers of science have been very clear on this point of Darwin's distinctiveness as an evolutionary thinker, since virtually every other evolutionist has held that science provided an incentive, if not outright epistemic basis, for a renewed assertion of faith. But insofar as professional scientists continue to believe that invoking 'Darwinism' in contemporary evolution debates is no more than a creationist ruse, they are spurning the admiration of some of their most learned fans.

In any case, if transhumanism is to mature into a proper political movement, it needs to drop the 'Darwin Pose' and take seriously that we are not simply one among many species, but are privileged by virtue of our capacity to understand the entire evolutionary process – indeed, courtesy of computer models, as if we had designed (but not 'determined') that process. This latter point explains the continued need for a deep (indeed, Abrahamic) theological orientation to any future biology. The very fact that we aspire to a 'science of life', many aspects of which we have already mastered for practical purposes (albeit not in some ultimate intellectual sense), obliges us to take responsibility for what we can now bring into (and take out of) the biosphere. Darwin would have had none of this. He was a precautionary thinker par excellence. So, while it is true that eugenicists – especially in Germany – were keen to invoke Darwin's name, they were mistaken to do so for the most part, and certainly with regard to eugenics' positive agenda of improving humanity that verges on contemporary transhumanism. For proactionaries, this is another reason to drive a wedge between Darwin and transhumanism.

In today's world, the real champions of transhumanism are not the casual pill-poppers and body-sculptors who simply want to extend their libertarian urges into new domains that will mainly benefit or harm themselves. If that were all there is to transhumanism, then the movement's many critics would be right to condemn its political shallowness. These lifestyle adventurers are literally 'figureheads' (i.e. ornaments on a ship's bow) vis-à-vis those who are engaged in the process of restructuring the governance of the planet, if not the universe, to realize species-level ambitions. Such restructuring will require not only fluency in manipulating the genetic code (*à la* 'synthetic biology'), but also clever political, economic and legal mechanisms for adjusting to the consequences of acting on those ambitions, not least by providing compensation for the failures – including deaths – that are bound to occur along the way. It is only this global normative reorientation that truly deserves the title of 'proactionary'. *The Proactionary Imperative* completes a trilogy (with Fuller 2011a, Fuller 2012) relating to the emergence of 'Humanity 2.0', that fork in human history which projects in both transhumanist and posthumanist directions. Whichever fork one takes, it will contribute to the opening move in a longer project of ideological reorientation.

The book consists of four chapters and concludes with a Manifesto that brings together all of our main points. Chapter 1 presents the emerging axial rotation of the ideological poles, effectively from *Left–Right* to *Up–Down*, with the latter duality exemplified by the 'proactionary' and 'precautionary' standpoints, respectively. We examine the theological, philosophical and scientific character of this radical political realignment as propaedeutic to an elaboration and defence of the proactionary principle as a progressive ideology, which constitutes the subsequent three chapters. Chapters 2 and 3 are meant to deal with complementary sides of what it means to assume the divine role of creator. Chapter 2 deals with theology, delving into what it means to take seriously our capacity to 'play God', for which we adopt the Greek term, *theomimesis*. Chapter 3 turns to the scientific and technological extension of our theomimetic capacity, focusing on the first explicitly proactionary science, eugenics, a field whose striking boldness of vision and failures in execution offer much insight for forging a future progressive ideology. Chapter 4 sketches out a proactionary legal framework, one that calls for a 'right to science' and specifically targets the collective ownership of genetic material, for which we introduce the term 'hedgenetics' (i.e. the genome treated as the basis for hedge fund investment).

Readers are entitled to know our intellectual starting points. Both of us are non-conformist Christians raised as Catholics. (Fuller calls himself a 'Unitarian', Lipińska a 'Deist'. The Enlightenment provenance of both views is not accidental.) This means that while we no longer defer to priests and rituals, we take seriously the biblical message that humans are created in the image and likeness of God and that Jesus is an exemplary case in point. Of course, Christians typically believe in a good many other things that may cause them to recoil at the positions that we discuss and advocate in these pages. But in our defence, it should be noted that we are keen to keep alive what distinguishes Christianity from the other world-religions, especially the non-Abrahamic ones. Many theologians nowadays treat the dominance of the scientific world-view as an opportunity to blur religious differences and even retreat from any claims of religion's cognitive significance. However, we believe that such 'ecumenism', the name most often attached to this adaptive strategy, does a disservice to the serious metaphysical differences between the world-religions, which correspond to equally sharp differences in science's role in the world (Fuller 2006: chap. 11).

In a nutshell, if you do not believe in the centrality of humanity to the cosmos – a distinctly Abrahamic theological preoccupation – then you will find the message of *The Proactionary Imperative* difficult to hear. However, if our book does speak to you, yet you see yourself as somehow poised 'against' or 'beyond' theology, then you need to question the source of your confidence in humanity's indefinite self-promotion. The historical evidence taken alone is far from edifying. Claims to 'human progress' are best read as adventures in 'pyramid scheme' finance, in which each successive generation buys into the promise that it will do at least as well as its predecessor, while those who sell the promise are really in the business of manufacturing believers in the scheme, so that they can at least break even on their own investment. The only non-cynical way to interpret this pyramid scheme logic is to suppose that if enough people believe that the promise will be redeemed, then one or more of these people will be sufficiently determined to deliver on it to everyone else's benefit. Thus, an enormous weight is placed on the efficacy of a self-fulfilling prophecy that once delivered can be presented as a 'triumph of the will'.

To be sure, no specific individual can be blamed for this scam – if that is what it turns out to have been. Yet, the mass improvement of public hygiene and medical care starting in the second half of the nineteenth century, alongside an inadequate sense of social planning for those who were thereby allowed to escape infant mortality, has meant that the world now contains an unprecedented number of people entitled to and expecting more than their parents can reasonably provide. One must have a strong sense of faith in humanity's capacity for self-transcendence to suppress the sense of callous irresponsibility that is *prima facie* suggested by this trajectory. Indeed, the authors believe that this history makes sense *only* if we take seriously that humans, understood either individually or collectively, are aspiring deities and that 'progressive' politics are dry runs at universal governance that in practice do not always go to plan. Thus, like more ordinary acts of faith, faith in science demands an active 'suspension of disbelief'. This is perhaps the best way to understand Karl Popper's falsifiability ethics of science: science can progress despite disappointment, setback and sheer error only because we treat these 'falsifications' as learning experiences rather than as proof that science itself has gone too far.

In contrast, Darwin refused to indulge this crypto-theistic narrative and so resisted the self-declared progressive causes of his contemporaries. We discuss this more in Chapter 3. As a result, we oppose 'Darwinism', though *not* the science that remains honorifically attached to his name, which has (fortunately) strayed quite far from his world-view. But even readers incapable of accepting our negative verdict on Darwin are challenged to provide a broadly compelling transhumanist narrative that does not implicate humanity's aspiration to god-like proportions. (Those daunted by this prospect should consider the rhetorical justification for republicanism in modern politics, not least in the founding of the United States of America, a nation that has remained inclusive of aliens without ever having completely abandoned its sense of divine inspiration.)

A central animating concern of *The Proactionary Imperative* that already appears in Chapter 1 is the ideological disarray of contemporary politics, which has only served to alienate the younger generation, despite the fact that the world is undergoing momentous socio-economic and technological changes. This alienation has been most damaging to the Left. The older style Leftists – call them 'Heritage Marxists' – are still rapturously received on university campuses, where the likes of David Harvey and Slavoj Žižek gamely trot out late nineteenth-century solutions to early twenty-first-century problems with the dutifulness of a Beatles tribute act. The newer style Leftists have gone from 'Red' to 'Green', adopting an exceptionally precautionary brand of politics that treats virtually every enduring human imprint on the planet as a source of fear and loathing. We were especially struck by the reactionary turn taken by the leading think-tank affiliated with the UK Green Party, when it called for a new parliamentary chamber with the power of veto over legislation that appears to compromise the freedom of future generations – as if such judgements could be epistemically or ethically well founded

(Read 2012). In contrast to all of these anaemic recent showings by the 'Left', the adolescent excesses of techno-libertarianism often on display in TED talks seem like a breath of fresh air.

This loss of faith in the progressive project among self-avowed 'Leftists' is perhaps most evident in the horror they express towards what Fuller (2010) has called 'Protscience', namely, the customized appropriation of scientific knowledge – typically via the internet and other academically unregulated media – as a basis for organizing one's life, typically in ways that scientists would find risky, to say the least. These range from those who pursue the Young Earth Creationist hypothesis that geological dating methods are radically wrong to those who follow the radically non-invasive medical regimes prescribed by homoeopaths. The term 'Protscience' alludes to the signature Protestant response to the authority of the Roman Catholic Church, which has been for believers to take the Bible into their own hands as a source of personal empowerment. In the process, they turn themselves into living laboratories of the faith, in which their interpretation of Scripture serves as hypotheses that demand to be tested.

Originally this prospect only frightened Catholics, but over the past two centuries this fear has been revisited upon Protestants – even liberal ones – who fail to appreciate Mormons, Christian Scientists and even Scientologists as kindred spirits (Bloom 1992). Put more simply, especially for those deaf to these religious resonances, it is as if today the Left has replaced the idea of science as a mode of inquiry or even a method with the idea of science as the final court of appeal on epistemic matters: National Academies of Science have become the new Vatican, with high-level peer review panels functioning as the Conclave of Cardinals. In contrast, proactionaries support the heterodox appropriation of science as epitomizing what in Chapter 4 we discuss in terms of a 'right to science' as a legal requirement to human self-realization. It leaves open the big question of how to harness all of this heterodoxy for collective benefit. However, this is not so different from the problem that faced Francis Bacon when, as lawyer to King James I, he sought to organize and evaluate the competing empirical insights of rival religious factions in seventeenth-century England. His solution, of course, was what we now call 'the scientific method'. We may now need a 'Scientific Method 2.0'.

Steve Fuller would like to thank the following people who provided valuable input, inspiration and opportunities for developing the arguments on these pages: Rachel Armstrong, Patrick Baert, Adam Briggle, Stephen Casper, Dylan Evans, Seyed-Morteza Hashemi Madani, Tomas Hellström, Britt Holbrook, Victoria Höög, Merle Jacob, Ian Jarvie, Pat Kane, Arie Kruglanski, Ed Lake, Luke Robert Mason, Andy Miah, David Budtz Pedersen, Project Syndicate, Rupert Read, Chris Renwick, Nico Stehr, Scott Stephens, Peter Thiel, Frank Tipler, René von Schomberg, Emilie Whitaker. Veronika Lipińska would like to thank Krzysztof Kormanski who supported, comforted and grounded her throughout the creative process. We thank Piotr Lipinski for introducing us to Michael Crichton's novel, *Next*. Finally, we would especially like to thank Sarah Chan for her very thorough and sympathetically critical reading of the manuscript, which spurred us to make some substantial corrections.

An earlier version of 'The Proactionary Manifesto' appeared as part of 'The World of Worth in the Transhuman Condition: Prolegomena to a Proactionary Sociology', a critique of Boltanski and Thévenot's *On Justification* (1991), published in Susen and Turner (2014).

1
Precautionary and Proactionary as the Twenty-first-century's Defining Ideological Polarity

1 Recalling the political theology of the old Right–Left divide

The modern Right-to-Left ideological spectrum is an artefact of the seating arrangements at the French National Assembly after the revolution of 1789. To the right of the Assembly's president sat the supporters of King and Church, while to the Left sat their opponents, whose only point of agreement was the need for institutional reform. The distinction capitalized on long-standing cultural associations of right- and left-handedness with, respectively, trust and suspicion – in this case, of the *status quo*. In retrospect, it is remarkable that this distinction managed to define partisan political allegiances for more than 200 years, absorbing both the great reactionary and radical movements of the nineteenth and twentieth centuries. But the decline in voter turnout in most of today's democracies suggests that this way of conceptualizing ideological differences may have become obsolete. Some have even argued that ideologies and parties are irrelevant in an increasingly fragmented political landscape. We strongly disagree. However, upon understanding what the old Right–Left division was really about, it becomes clear that it is now due for a 90-degrees rotation on its axis to recapture the spirit of the original division. That spirit is defined in a question: *should it be presumed that the past dictates the future, unless proven otherwise?* Those on the Right say 'yes' and hence practise a positive politics of induction; those on the Left say 'no' and hence practise a negative politics of induction. It is against this backdrop that we propose *precautionary* and *proactionary* as the poles defining, respectively, the 'New Right' and the 'New Left'.

Nowadays it is common to construct the ideological spectrum by placing conservatives on the Right, liberals in the middle and socialists on the Left. The resulting pattern leaves the impression that the metaphysical individualism associated with liberalism anchors the spectrum, with the extreme ends on both sides occupied by collectivists who base group identity on either family or race (the Right) or class or state (the Left). However, this default interpretation, while perhaps correct in some of the detail, is clearly not true to the spirit of 1789. In the original National Assembly, as just mentioned, the centre was occupied by the *status quo,* and the question dividing the two sides was whether society should re-dedicate itself to the historic roots of the *status quo* (which had become corrupt in the recent past) or break decisively with the past in search of a more forward sense of self-legitimation. It was in this context that the people who would soon be known as 'reactionaries' sat on the right of the conservatives, while the people whom we would now consider 'liberals' and 'socialists' sat together on the left.

Over time, and for reasons that will be explored below, liberals and socialists increasingly parted company – but still in alternative ways of breaking with the *status quo*. Generally speaking, liberals would have people face the future as individual agents from whose aggregate decisions emerge an overall sense of direction for society, be it defined politically in terms of majority rule or economically in terms of dominant market share. In contrast, socialists would have people face the future as one collective agent explicitly dedicated to such a specific direction. Thus, liberals stress 'equality of opportunity' and socialists 'equality of outcome', both understanding a trade-off between the two forms of equality. But on the negative side, for the socialist, 'equality of opportunity' means that some will race ahead, while for the liberal 'equality of outcome' means that some will be held back. Liberals take the difference between 'progressive' and 'reactionary' as always in flux, with votes and prices signalling changes in direction, while socialists see the difference as institutionalized in a more principled way, as electoral defeats are replaced by purges and market failures by expropriation. To put it in the metaphysical language popularized by Michael Dummett (1977), liberals are *anti-realists* and socialists are *realists* about the future.

Unlike their right-of-centre colleagues, liberals and socialists agree that the future – not the past – provides the ground for societal legitimation. But that is still not quite the right way to distinguish the ends of the ideological spectrum. In particular, what distinguishes liberals and socialists with regard to the future is their rather different attitudes towards the past – especially when the past has not turned out as they would have liked. While it would seem natural to interpret the 1789 Right–Left split in terms of a past versus future orientation, in fact *all* the ideologies looked to the past in one crucial respect: for an appropriate account of human nature – specifically, of human potential. However, they differed in terms of how much of that potential has been revealed in actual human history. The right-wingers believed that most or all of that potential had been already revealed, such that long-surviving patterns of conduct were the ones worth taking forward into the future. (In this respect, despite his belief in the ultimate triumph of liberal democracy, Francis Fukuyama [1992] remained very much a student of the ultra-conservative Leo Strauss in his confidence that history has already revealed the full range of feasible polities.) The Left-wingers held that relatively little of that potential had been realized, but substantially new social arrangements would provide the opportunity to reverse that state-of-affairs. True to Bismarck's definition of politics as the art of the possible, behind this difference in sensibility lay alternative metaphysical interpretations of what is 'possible'.

Right-wingers clung to an understanding of what is possible that would have been familiar to Aristotle and remained largely unchallenged until the Franciscan scourge of Thomas Aquinas, John Duns Scotus, took to the world-historic stage in the fourteenth century. Aristotle had effectively equated the possible with the empirically probable, itself a gloss on 'natural'. In contrast, Left-wingers availed themselves of Duns Scotus' more modern 'semantic' identification of the possible with the conceivable – that is, including logically coherent yet unrealized states-of-affairs. In shifting the meaning of the possible from what has been experienced to what might be realized, Duns Scotus had effectively elevated humanity from the highest animal to an aspiring deity. That is the moment when 'possible' began to stand for a 'Left'. After Scotus, we were no longer creatures whose existential horizons are defined by the collective experience of our ancestors but rather ones who within each generation possesses the capacity to construct the world anew from first principles – as God himself did, in Augustine's phrase, *creatio ex nihilo* (Fuller 2011a: chap. 2).

One downstream effect especially relevant to the proactionary principle is worth flagging here, as it will be stressed later in the book: namely, the secular reinvention of divine creativity that occurred five centuries after Scotus when another clerical follower of Augustine, the abbot Gregor Mendel, discovered that the factors responsible for the hereditary transmission of traits do not simply draw upon the actual history of sexual reproduction but are constituted as a set of permanent possibilities that might be realized at any time under the right conditions. Thus, the science of genetics emerged in the early twentieth century to determine how that might happen in practice, a task that has turned out to be trickier than first supposed. In any case, true to the proactionary spirit, theology and biology become one in the aptly named field of *genetics*.

This historical trajectory also explains the emphasis that we place in this book on 'eugenics' as a project that, if nothing else, was intended to enable humans to step up to their divine capacity by taking responsibility for the successive production – as opposed to mere 'reproduction' – of nature. By the time Francis Galton coined 'eugenics' in the late nineteenth century, humans had domesticated many animal and plant species – not to mention the physical environment more generally – to great effect. But our self-domestication remained shrouded in the pseudo-sociology of inheritance law, whereby capital was prescribed to travel along lines of familial descent. In this respect, eugenics marked a 'coming of age' for humanity when it started to take seriously its powers over life and death that it had taken for granted for centuries. We shall return to this point in Chapter 3.

In our own day, the Scotist revolution has not escaped critical notice by those comprehensively conservative religious thinkers who call for a 'neo-orthodox' revival in Christianity (Milbank 1990). In this context, Duns Scotus stands accused of having combined and radicalized two strands in Augustinian theology: (a) God is (always) free to create any conceivable world; (b) we are created in the image and likeness of God. From these premises it is then easy to conclude that we have an obligation to explore those unrealized possibilities (Funkenstein 1986: chap. 2). In that case, the fact that in 1789 France the established church continued to support the *status quo* – a hereditary monarchy, even after it had been shown to be corrupt – appeared as an affront to those who believed that our divine entitlement rendered us capable of much more than simply perpetuating the legacy of previous generations. Indeed, humans may have the wherewithal to constitute a government from first principles, the sort of 'second creation' adumbrated in eighteenth-century social contract theory that had been put into practice on a large scale only a few years earlier in the founding of the United States of America (Commager 1977).

This Scotist mentality, which marked where the Left broke most sharply with the Right in the French National Assembly, is characteristic of what we are calling the 'proactionary' pole of the newly emerging ideological spectrum. In effect, it interprets the 'meek' in the third verse of Jesus' 'Sermon on the Mount' – 'Blessed are the meek; for they shall inherit the Earth' (Matthew 5:5) – to refer to humanity's unrealized potential to rule itself (despite its current state of powerlessness). Much of the 'prophetic' strain in modern evangelical Christianity stems from this interpretation (Swartz 2012). In contemporary US political philosophy, its subtlest advocate has been Roberto Unger, its most popular one Cornel West. The two teamed up in Unger and West (1998).

The most general practical consequence of the Scotist shift was that people came to take the better worlds that they could imagine as no mere passing fantasies but as motivators to action. This fundamental change of attitude to the contents of one's own mind came to be widely held only in the second half of the eighteenth century and may have been related to the concentrated doses of alcohol and caffeine that started to circulate in European brains (Fuller 2012: Epilogue). However, the shift can also be clearly noted in the previous two centuries, albeit esoterically, in the striking tendency of the founders of the Scientific Revolution – most notably Johannes Kepler – to promote the cognitive significance of their dreams from not simply predictions, as *per* the ancients, but outright blueprints for understanding reality (Koestler 1959).

In the run-up to the Scientific Revolution, Duns Scotus' radical reinterpretation of 'the possible' was popularized by John Wycliffe, who rendered his teacher's revisionary scholasticism concrete by having the Bible translated into English so as to unleash human potential. This project finally received royal approval more than two centuries later with the publication of the King James Version in the early seventeenth century. The King's lawyer, Francis Bacon, shared this spirit as concomitant with the experimental method, which he famously portrayed as extracting from nature secrets that it might otherwise hide forever (Fuller 2008a: chap. 2). While much has been made of the suspicion if not outright hostility towards nature that is reflected in Bacon's sentiment, it is perhaps best understood as humans seeing in nature what they regard as being in most need of correction or elaboration in themselves, given the hereditary burden of Original Sin that attaches to our animal nature (Harrison 2007).

Duns Scotus had paved the way linguistically for Bacon's vision, which was now proposed to harness the new science to the political ascendancy of England, by introducing a manner of speaking that analytically detached God's attributes (i.e. omnipotence, omniscience, omnibenevolence) from a unique deity. Scotus' linguistic innovation made it possible for humans to aspire to god-like powers without outright turning into God, thereby staying on the right side of religious heresy (Brague 2007: chap. 14). Of course, theists had to entertain an increasingly problematic – and ultimately secularizing – consequence of the Scotist move: namely, that divine attributes differ from corresponding human ones only by degree and not kind, which in turn has been the basis for *both* the 'literalist' reading of the Bible and the idea that nature can be read as a book written in a decipherable (typically mathematical) code (Fuller 2010: chap. 5). In any case, the subtle but systematic abstraction of divine function from divine substance begun by Scotus unleashed enormous consequences ranging across logic, physics and economics, resulting in a conception of value based on efficient exchanges of energy, as humans tried to approximate God's capacity to create *ex nihilo* (Cassirer 1923; for more critical views of the same development, see Mirowski 1989, Rabinbach 1990). We return to this theme in the next chapter under the US engineer Buckminster Fuller's (1968) rubric of 'ephemeralization'.

One complicating factor in defining the original Right–Left divide was the emergence of comparative cross-cultural histories of governance in the half-century prior to the French Revolution, most impressively by Baron de Montesquieu. Officially presented as updating a line of inquiry initiated by Aristotle, both a 'Right' and a 'Left' spin was given to its eighteenth-century revival. Right-wingers (e.g. David Hume) concluded that the variety of governance patterns found throughout the world argued against the possibility of a universally applicable blueprint for social organization. After all, each society, true to the accumulated experience of generations of its members inhabiting the same place, would have hit upon custom-made social arrangements. In the nineteenth century, ideologies that we now recognize as both 'cultural relativist' and 'racist' – often not clearly distinguished from each other – developed this approach, typically to promote a conception of the state based on 'nationality'. In contrast, Left-wingers (e.g. Marquis de Condorcet) interpreted the variety of governance patterns as alternative realizations of a universal human potential, from which everyone may learn as we converge on a common progressive trajectory. Implicit here is the prospect that humanity is collectively advanced by tapping into opportunities already present in some culture's past but which have yet to be fully realized or sufficiently extended (Fuller 2011a: chap. 1). Perhaps ironically, a latter-day descendant of this sensibility may be found in the attitude of transnational pharmaceutical firms, armed with medical anthropologists, who see the world's cultural diversity as so many simultaneous experiments in collective human survival (Brown 2003: chap. 4). Nevertheless, we believe that overall Condorcet has the better side of the argument.

2 Right vs. Left as a contest over the past to determine the future

As we have just seen, the original Right- and Left-wingers were arguing from much the same empirical base, but whereas the right-wingers treated the sheer survival of social practices as self-validating and hence stressed the costs of deviating from them, the Left-wingers conjured the benefits that would have been (and perhaps may still be) accrued by pursuing versions of known alternative practices. This difference may be seen as a political version of the complementary relations exhibited by matter in motion at the quantum level that Werner Heisenberg formulated as the 'uncertainty principle': *The Right espouses a politics of position, the Left a politics of momentum.* The Right holds that we are where we belong, while the Left presumes that where we are is no more than a state in motion. At stake here is what the analytic philosopher Nelson Goodman (1955) originally called 'projectibility', which he described as the 'new riddle of induction' – in short, which aspects of the past are worth projecting into the future? (Goodman himself imagined two predicates, 'grue' and 'green', both of which are true of all emeralds before now but 'grue' claims that in the future they will be blue not green.) The original 1789 ideological divide vividly illustrates why the answer is far from obvious – though in less dramatic ways judges routinely face a version of this problem when selecting cases as precedents for framing the case under adjudication.

On the one hand, the Right-wingers practise a kind of 'straight rule' induction, whose presumption is that the future continues the dominant tendency in two senses of 'dominant': given our knowledge of the past, it is the most obvious course of action in light of the most obvious framing of the situation. Thus, special reasons must be offered to change a course of action that has been established on, respectively, such empirical and conceptual grounds (cf. Fuller and Collier 2004: chap. 10). This general approach, admitted by Hume to be our default habit of mind, is properly called 'conservative'. It was accorded a metaphysically (and politically) elevated status as the working of 'natural reason' by the cleric Richard Whately (1963) in the most authoritative logic textbook in early nineteenth-century Britain.

On the other hand, the Left-wingers interpret the dominant tendency as an extended contingency that is reversible under the right conditions to reveal alternative lines of thought and action that had been obscured or suppressed. The difference between liberals and socialists on this score has turned on whether any of those alternatives are, so to speak, 'The Truth-in-Exile'. Generally speaking, liberals say no, socialists say yes. Whereas liberals hold that any alternative is in principle realizable under the right circumstances, socialists privilege a limited number – if not simply one – of those alternatives as providing an authentic realization of human potential (of course, without denying the need to apply force to enable its realization). Thus, while liberals have focused on maintaining an ever-present capacity to reverse any regime that happens to be dominant at the moment (e.g. via regular elections, free markets), socialists have concentrated on identifying the one true regime that is worth pursuing in the face of anticipated resistance, as it overturns entrenched habits of thought and action.

Informing this division in the Left is the dual character of the deity implied by the Scotist revision of the concept of possibility previously mentioned. God is the only being who can do whatever he wants and whatever he does is what he wants. (The hidden premise is that the deity's 'wants' are 'oughts', by definition of the deity's supremacy.) The former clause captures the liberal's and the latter the socialist's aspiration for humanity in light of our having been created *in imago dei*. From these alternative theological spins, flow opposing conceptions of justice. For liberals, justice is a matter of procedural fair play, whatever the outcomes, whereas for socialists it is a matter of reaching the right result, perhaps by whatever means. Rawls' (1971) method of 'reflective equilibrium' may be seen as an attempt to reconcile these competing intuitions – 'justice of the means' and 'justice of the ends', so to speak – in service of a transcendental argument for the welfare state.

At a still deeper level lies a difference of metaphysical interpretation – specifically, of the 'human potential' that both liberals and socialists accuse right-wingers of short-changing. Here it is useful to recall the distinction between two Hegel-inspired concepts: Freud's *sublimation* and Marx's more faithful conception of *sublation*. Sublimation implies that (libidinal) energy passes through many forms without ever quite losing its original character, whereas sublation implies a more fundamental transformation that can only be fully understood once (labour-power) energy reaches its final state of organization. The former captures the liberal's sense of the body politic's momentum, the latter the socialist's. From this standpoint, a truly liberal account of capitalism is Max Weber's *The Protestant Ethic and the Spirit of Capitalism*, the latter phrase in the title is understood as a sublimation of the former phrase. Further sublimation transpires in the twentieth century as Protestantism's self-transcending productivist impulse migrates from the manufacture of consumer goods to the manufacture of one's own sense identity through what Thorstein Veblen memorably called 'conspicuous consumption'.

Karl Popper (1957) notoriously got the epistemic measure of the difference between liberals and socialists in terms of two senses of 'expectation' that reflect different attitudes that liberals and socialists have towards the future: *prediction* and *prophecy* – the former the cornerstone of the scientific method (*qua* Popper's own falsifiability principle) and the latter the utopian hope that fuels radical politics, both sacred and secular. Thus, the 'prediction' pole belongs to the piecemeal social engineers, whom Popper prefers, and the 'prophecy' pole to the revolutionaries who justify their policies in terms of historical destiny. On the one hand, Popper's social engineers aim to keep politics maximally open to new possibilities by ensuring that any course of action taken is reversible in light of the consequences. On the other hand, his dreaded revolutionaries are keen to eliminate alternative possibilities for action that might divert society from reaching its ideal state. However, despite their stark differences, both predictors and prophets are positively disposed towards the future, especially the long run. Moreover, both provide mental preparation for particular disappointments along the way – the predictors anticipate corrigible error (typecast as 'ignorance'), while the prophets anticipate surmountable obstacles (typecast as 'enemies').

In its day, Popper's notoriety came from challenging the scientific credentials of Marxist 'historical' or 'dialectical' materialism – indeed, by turning the Marxist standpoint on its head, arguing that the very liberals whom Marxists despised (under such epithets as 'idealist', 'Machian', 'positivist') practised a truly scientific politics because they submitted their knowledge claims to fair tests, be it in the ballot box or the marketplace. Here Popper took a page from Max Weber's original stereotyping of the open-minded scientist and goal-oriented politician, as portrayed in the two great speeches of his later career, 'Science as a Vocation' and 'Politics as a Vocation'. Weber's contrasting presentation of how scientists and politicians coped with failure mapped onto the more general action-orientations, respectively, *Wertrationalität* ('value-rationality'), which covered both scientific and religious practices, and the *Zweckrationalität* ('goal-rationality'), which covered both political and economic practices.

However, the dichotomization is simplistic. According to the Weberian stereotype, when faced with failure, the scientist switches hypotheses while the politician carries on as if nothing had happened. But here it is important to compare like with like. After all, the scientist seeks truth with the single-mindedness of a politician who seeks power. For example, I may favour elections as a means of selecting leaders either because elections force people to think about leadership in the right way (i.e. *wertrational*) or elections are an efficient means to come up with the right leader (i.e. *zweckrational*). The former would lead me to extol campaigning and voting as expressions of civic virtue, an intrinsic political good regardless of who actually got elected, while the latter would lead me to think about more efficient means of achieving the aim of effective leadership, which may include so-called strategic voting (i.e. voting, if at all, for someone other than your preferred candidate). Similarly, I may uphold Popper's criterion of falsifiability either because it forces scientists to think about their hypotheses in an appropriately critical-rational frame of mind (i.e. *wertrational*) or it does the best job of getting scientists closer to the truth (i.e. *zweckrational*). The former would lead me to focus on embedding falsifiability into the scientific culture, whereas the latter would lead to me to seek more efficient versions, if not outright substitutes, of falsifiability.

But the matter can be approached with still greater subtlety: falsification does not demand that the scientist give up the overall direction of her inquiry once her hypothesis is shown to be false – that is, she does not abandon her motivating metaphysical world-view, which extends well beyond what can be justified simply in terms of a discipline-based Kuhnian paradigm (Agassi 1975). Rather, the falsificationist concedes that realizing the sort of world anticipated by her metaphysics inevitably requires pursuing a different line of empirical inquiry, one that incorporates elements of her previous pursuit but now re-oriented towards different specific outcomes. Thus, the post-mortem of a falsified hypothesis involves not simply avoiding a class of untenable predictions in the future, but more importantly incorporating the error as a guide to building a richer theory that then provides the basis for new hypotheses (cf. Hegelian sublation) – as opposed to an *ad hoc* repair that would allow the theory to advance as if nothing had happened. To insist on the theory's abandonment would effectively deny the information value of the falsification through forced extermination of its content.

All of this is not so very different from a politician who is flexible with regard to tactics while pursuing a strategy whose constancy is not deterred by specific setbacks. Perhaps the key difference is that the politician would aim to publicize only the self-fulfilling – and not the self-defeating – consequences of her strategy. While the public admission of error is seen as a mark of integrity in a scientist, it is often taken to be a mark of incompetence in politician. (However, popular histories of both science and politics tend towards the self-serving concealment of all but the most instructive failures; hence, the application of the term 'Whig' to both sorts of histories: cf. Brush 1975.) Nevertheless, scientists and politicians may learn equally well from error, even as the latter fail to say so openly. In this context, it is worth recalling the high esteem in which Enlightenment politicians, not least the US founding fathers, held *hypocrisy*, a state of divided consciousness that requires the politician to be sufficiently confident in his own ultimate right-mindedness to self-justify various reversals of fortune without admitting them publicly (Runciman 2008). The closest that science comes to admitting the value of hypocrisy may be Popper's (1972) own strong distinction between the *beliefs* and the *theories* held by the scientist: Popper does not care what beliefs scientists (privately) hold as long as they hold their theories (publicly) accountable to evidential tests (Fuller 2007b: chap. 3).

In the history of the philosophy of science, this strong distinction between one's personal beliefs and theoretical assertions is normally associated with 'instrumentalism', a position popularized by the logical positivists, who reduced the content of scientific theory to the evidence that supports it – in that sense, a suitably operationalized theory is no more than a machine for generating evidence. However, instrumentalism emerged a little over a century ago from the Roman Catholic physicist Pierre Duhem (1969). Duhem had been deeply influenced by the then-recent opening of the Vatican archives to the records of the trial of Galileo, in which the difference between what was directly evidenced and what could be inferred only given prior beliefs was very much at play. The lesson that Duhem drew was that for *both* Galileo and his Jesuit Inquisitors, faith in God provided an unerring but not directly scrutable guide for their inquiries. Nevertheless, by trying to cash out this belief in agreed terms of evidence (say, as the outcome of an experiment), each managed to keep alive their respective beliefs, despite the inevitable empirical setbacks, and in a way that could inform both sides. Such a lesson proved especially useful in the secular political environment of Duhem's own Third Republic France, where instrumentalism functioned as a brake on 'scientism', the steering of science for specific political ends. (A Duhemian for our times is Bas van Fraassen [1980].)

However, Duhem's epistemic grounds for, so to speak, 'scientific hypocrisy' could not be more different from Popper's: Duhem kept his theism private to protect its capacity to illuminate scientific inquiry in the face of freely chosen theories that every so often are subject to overextension and falsification, whereas Popper was more concerned that privately held beliefs with no clear criteria of public testability did not contaminate the course of scientific inquiry. For Duhem the hypocrisy embodied in science's belief-masking technical discourse and laboratory rituals was insurance against scepticism and the abuse of science by the dominant political party; for Popper hypocrisy insured against relativism as well as the pressure towards consensus within the science itself. However, neither Duhem nor Popper realized that hypocrisy might be what Jon Elster (1998) has called, with a nod to Benjamin Franklin, a 'civilizing force' on science (Fuller 2000: chap. 8; Fuller 2009: chap. 4). In other words, even if one's personal beliefs remain hidden, one's prolonged engagement in public life – be it as a politician or a scientist – may unwittingly serve to alter those beliefs over time, if only to minimize any sense of cognitive dissonance between one's private and public faces. This phenomenon is familiar to social psychologists as *adaptive preference formation*, but its exact interpretation is contestable. It is often pejoratively equated with 'rationalization'. However, in this case, it may be understood more neutrally – if not positively – to imply that by doing science, you come to want the sort of beliefs that science is likely to validate over time, which in turns shapes the character of your putatively non-scientific views. Dissenting Christian theologies from the Enlightenment, such as Deism and Unitarianism, count as adaptive preferences in this sense.

In the 1950s, Leon Festinger and his colleagues at Stanford developed the concept of adaptive preferences to explain how a religious sect that falsely predicted the end of the world managed to carry on preaching its gospel (Festinger et al. 1956). Their work left the impression that the sect had developed a defence mechanism, 'sweet lemons' as Elster (1983) memorably called it (the converse of 'sour grapes'), which allowed them to cope with the falsification with minimal adjustment to their core beliefs. However, closer attention to the details of the sect's behaviour suggests that its members engaged in what metaphysicians call a 'modal' analysis of their beliefs – that is, the sect interrogated what remained possible, impossible, necessary and contingent within their belief system after the falsification. They ended up attributing their epistemic failure to features of their beliefs that were not necessary to hold for purposes to advancing their cause, while at the same time explaining better (at least to their own satisfaction) their own understanding of God's word.

To be sure, the sect's autocritique did not appease its opponents, who would have simply liked the sect to disappear. However, it did serve to bring the sect's epistemic standards in alignment with those of other faith communities. In effect, the modal analysis generated intellectual antibodies that strengthened the immunity of the sect's belief system to radical external challenge. There may be a more general epistemic lesson here that plays into the 'proactionary' pole of the emerging ideological spectrum. Whereas Popper used to identify humanity's evolutionary advantage in terms of our capacity for our theories to die in our stead, stressing the distance between our conceptions and our selves, he meant that our tolerance for a theory's death reflects our capacity to incorporate the theory's living aspect (cf. Fuller 2007b: chap. 3). It gives new meaning to Nietzsche's Zarathustrian maxim: 'What doesn't kill me makes me stronger'. In short, our resolve to become *Homo scientificus* is strengthened with every false scientific theory we embrace, reject and move forward from. In more vivid theological terms, when the New Testament presents the 'resurrected' Jesus as someone who has returned from the dead with an improved body, he should be seen as having retained (if not clawed back) what truly matters from the toll levied by the moment of death. In this respect, 'progress' in a way that Nietzsche, Popper and proactionaries could equally recognize is about perpetual recovery (or re-booting) in the face of imminent extinction.

3 Precautionary and proactionary as the new polar principles

One ideological division can re-invent the Right–Left distinction for the twenty-first century – albeit in a new key: *precautionary* versus *proactionary* attitudes towards risk as principles of policymaking. To be sure, the two principles are not yet on an equal footing. On the one hand the precautionary principle increasingly figures in environmental and health legislation. It is normally understood as the Hippocratic Oath applied to the global ecology: Above all, do no harm. One familiar precautionary measure is the policy of reducing human reproduction as a means of reducing our carbon footprint on the planet: even if it does not resolve the current ecological crisis, it will slow down its effects. On the other hand, the proactionary principle is the eighteenth century Enlightenment idea of progress on overdrive. Proactionaries often write as if science and technology have historically charted a path of unmitigated success. For them all that inhibits the indefinite extension of our distinctly human powers is ignorance and fear, both of which are seen as remediable with greater knowledge. Proactionaries give little credence to the concern that our increasingly high-tech *and* high-risk world points to a fundamental misapprehension of the human condition. As of July 2013, Google hits for 'precautionary principle' outnumbered those for 'proactionary principle' by more than 50 to 1.

The precautionary–proactionary distinction can be encapsulated in social psychological terms as a difference in 'regulatory focus': precautionary policymakers aim to prevent the worst possible outcomes, proactionary ones to promote the best available opportunities (Higgins 1997). Metaphysically speaking, the difference boils down to the management of modality: on the one hand, precautionaries draw a very sharp distinction between the actual world and other possible worlds – an actual loss can never be compensated by the possibilities that are thereby kept open. For precautionaries, the value lost through species extinctions cannot be offset by however much room is thereby left to humans to expand their lives. On the other, proactionaries are quite open about their willingness to sacrifice a significant part of present-day conditions to enable the future to stay open – for them, even when things go horribly wrong, it is less an outright loss than a learning experience. In short, whereas precautionaries regard significant risk-taking as ultimately corrosive to our freedom, the limits of which are already evidenced in the actual world, proactionaries regard risk-taking as necessary to discover the limits of what is possible, which by no means is exhausted by what has already happened.

The precautionary principle began life in early nineteenth-century Germany as *Vorsorgeprinzip*, as Georg Ludwig Hartig was laying the scientific foundations for forestry. For Hartig, whose name nowadays graces a leading German charity dedicated to environmental sustainability, the precautionary principle entailed that each generation should leave the next one with forests in the same state in which they found them (through a policy of conscientious re-planting of cut down trees, etc.). This formulation of the principle persists to this day in a much more generalized form, often featuring in Green Party proposals for defining just governance in terms of enabling future generations to live lives at least as fulfilling as our own (e.g. Read 2012).

However, the precautionary principle's origin in forestry highlights its contestable normative assumptions, including these two: (a) a steady-state (i.e. no net loss *or growth*) approach to *both* human and forest replacement; (b) a denial that the needs and wants currently satisfied by forests might be satisfied by something else (perhaps entirely artificial) in the future. Whatever one makes of these assumptions, applied either locally or globally, they derive their normative force from a sense of nature's ultimacy that precedes or supersedes human ingenuity. Indeed, this is why the United States insisted on characterizing precautionary as an *approach* rather than as a *principle* in the 1992 Rio Declaration on Environment and Development, as the Americans thought the latter would have surreptitiously introduced a sense of natural law that was inappropriate to international ecological policy negotiations (Garcia 1996).

From the standpoint of the history of economics, the logic informing the precautionary principle resembles less that of modern capitalism than of its eighteenth-century predecessor, physiocracy. The physiocrats, mostly French Enlightenment philosophers, tied productive capacity directly to the material character of the economic inputs – say, the number of trees and humans – rather than to their effective output – say, the value derived from a given number of trees or humans, which may (in principle at least) be produced by some other means more efficiently, and perhaps even in the absence of the original trees or humans. Indeed, the near-magical character of 'labour' as a source of value in classical political economy from Smith and Ricardo to Mill and Marx lay in just this capacity to transform one form of capital into a more efficient form, which obviates the need to resort to the precautionary principle's steady-state thinking – or its updated, somewhat more liberalized versions, 'sustainability' and 'carrying capacity' (Jacob 1996). However, classical political economy suffered from two blind spots concerning the development of capitalism – only one of which even Marx anticipated – that contribute to the precautionary principle's continued relevance today.

The first, partly anticipated by Marx, is the relative ease with which natural forms of capital would be replaced by artificial forms, not least including the mass replacement of human by machine labour, which in turn has periodically fuelled thoughts that the human body itself might be surplus to requirements in an optimally efficient economy – which is to say, one that is fully technologized. In that case, what both the physiocrats and today's precautionaries would take as the inviolate source of value – namely, 'nature' in its normal embodiments – may come to be treated under the logic of capitalism as disposable waste. In this important sense, capitalism, despite its reputation for being 'materialistic', is much less respectful of embodied nature than earlier economic systems, which typically included ecological stewardship in their remit. On the contrary, the popular nineteenth-century scientific idea that matter was ultimately a container of 'energy' waiting to be harnessed for maximum productivity served to re-invent in a materialist key classic spiritualist values such as asceticism – now rendered as 'efficiency' (Rabinbach 1990). From this standpoint, what Friedrich Engels (1939) called the 'dialectics of nature', 'materialism' and 'idealism', are polar philosophies fixated on distinct moments in the history of 'energy', the former focused on the raw material that has been inherited and the latter on its more efficient deployment in the future.

However, Marx did not foresee the second blind spot, which is that the ingenuity of human labour would result in the manufacture of not only new products that satisfy current human needs more efficiently but also new human needs that then demand new products. In short, classical political economy underestimated the significance of advertising in allowing for the relatively peaceful 'anticipatory governance' of consumption, as producers sought to open up new markets once the old ones have been saturated. (Indeed, Marxists thought, on the contrary, that the inevitable saturation of domestic markets would force producers overseas, eventuating in a succession of imperial wars.) More specifically, as the permeation of the 'cash nexus' injected exchange value relations into more traditional sources of social meaning, one's sense of identity – and increasingly individuality – came to be something the continual maintenance and upgrading of which one took personal responsibility for. When Max Weber's great rival Werner Sombart first used 'capitalism' in a book title in 1902, it was to this transformation that he referred (Grundmann and Stehr 2001).

More than a century later, the result is that we are awash in products whose threat to the global environment offsets any efficiency gains that have been made in their production. Although, as we shall see below, proactionaries can counter the more moralistic versions of this critique of 'consumerism', the precautionary sting remains in the prospect that increased productivity will never adequately recover the costs of increased production. In the end, entropy conquers all. A first attempt at a proactionary response to this gloomy prognosis has appeared in the so-called *Hartwell Paper* drafted by several eminent economists and social scientists who do not dispute the fact of significant short-to-medium term climate change but treat it as offering an unprecedented opportunity for innovative energy investments (LSE Mackinder Programme 2010).

For its part, the 'proactionary principle' first appeared under that name as the title of a declaration drafted by the transhumanist philosopher Max More (2005) and agreed by a congress of like-minded thinkers – including such champions of indefinite human longevity as Ray Kurzweil and Aubrey de Grey – at the 2004 'Progress Summit', sponsored by the Extropy Institute of Austin, Texas. The principle was explicitly designed as a foil to the precautionary principle. More specifically, 'The Proactionary Principle' was occasioned by the appearance of George W. Bush's Bioethics Council Report, which *inter alia* invoked 'natural law' to call for a ban on US federal funding of stem cell research (Extropy Institute 2004). The Report observed that the technology requires the sacrifice of many embryos in a largely trial-and-error process, which even when successful cannot guarantee that the generated organs will perform as desired. Thus, once the speculative nature of stem cell research's life-enhancing potential was set alongside the known destructive character of such research in practice, the Council concluded that a ban was required. In contrast, for proactionaries much greater long-term political and economic risks are assumed by *not* pursuing stem cell research, given an already growing population living into old age but in a condition that places an increasing burden on healthcare and welfare provision (Fuller 2011a: chap. 3). From that standpoint, stem cell research represents the entry point into what Princeton molecular biologist and avowed proactionary Lee Silver (1997) has called 'reprogenetics', a technology capable – at least in theory – of producing functioning organs ('spare parts') on demand, thereby providing an important platform for launching a credible programme of healthy indefinite life extension.

'The Proactionary Principle' was perhaps most innovative in associating this ban with the politics of the precautionary principle. As the appeal to natural law above suggests, Bush's Bioethics Council was populated by what transhumanists like to call 'bioconservatives', including several clerics, who adopt a broadly Aristotelian moral horizon that stresses the necessary rootedness of social convention in 'natural' attitudes and responses to the world (Kass 1997, Fukuyama 2002). Here 'Aristotelian' refers to an emphasis on 'human nature' as both a biological and a normative entity that requires specific sorts of environments for its full realization. Informing this sensibility is a more general metaphysical belief that there are objectively 'natural' and 'unnatural' ways of being human, for which history is our most reliable source. Some bioethicists have come to deride this sensibility as exhibiting the 'yuck factor', given the enormous moral significance that bioconservatives seem to grant expressions of disgust, be it explained as a spontaneous moral sentiment or an evolutionary survival strategy (Briggle 2010). Indeed, the two explanations may be paired together, as in the case of the Leo Strauss student, Larry Arnhart (1998), who persuasively shows the natural fit between Darwin's own world-view and the bioconservative perspective.

To be sure, the Aristotelians are not the most obvious bedfellows of the eco-friendly, species-egalitarian types who champion the precautionary principle and think of themselves as occupying the Left of the political spectrum, perhaps even to the left of mainstream socialist parties. Nevertheless, beneath these surface political differences rests agreement on a sense of 'nature' that pre-exists or transcends human activity and which sets significant limits on what humans can ever hope to accomplish. Here Peter Singer (1975) deserves credit – or blame – for the syncretistic feat of transferring the political concern for 'equality' from a first-order relationship among human individuals to a second-order relationship among species, even though Aristotle's own species-centred approach to biology implied a hierarchy of species souls that had been already abandoned by Jean-Baptiste Lamarck two centuries ago and, more to the point, by Darwin 50 years later (Fuller 2006: chap. 13). Nevertheless, the result has been an unwitting alliance of conservative Christians, ecologists and communitarians all wishing to tie our sense of humanity to the recognition of natural limits, which may be glossed as a fall from divine grace, our animal mortality or, more simply, the sheer finitude of our distinctly human powers. The latent misanthropy of this position is revealed in the popular eco-euphemism, 'anthropocene', which refers to the ongoing collective impression that our species leaves on the planet, as if the meaning of humanity were reducible to its carbon footprint rather than the journey that is evidenced in such ecological stigmata (Crutzen 2002).

One current political theorist whose world-view borrows from both natural law tradition and more modern communitarian and ecological thought epitomizes the new precautionary ideologue. I mean that indefatigable foe of perfectionists and utilitarians more generally, the Harvard political philosopher Michael Sandel (2007, 2012). For Sandel, social life possesses intrinsic virtues that have been developed through traditions of practice, which are compromised if standards borrowed from science, technology or the market are used to reform them. Thus, what confers value on a well-lived human existence cannot be reduced to a cost–benefit matrix. While Sandel may be now in the limelight, other theorists with longer and more distinguished track records are tilting in the same direction. For example, there is the 'capabilities' approach to 'quality of life' promoted by the welfare economist Amartya Sen and the virtue theorist Martha Nussbaum (Nussbaum and Sen 1993). This shifts the justification of the welfare state away from the rational assent of its potential members – as in John Rawls' *A Theory of Justice* (1971) – and offers instead a social scientist's inventory of the necessary attributes of a decent human life, regardless of assent. Finally, there is the great German sociologist Jürgen Habermas (2002), who has come to share the Roman Catholic Church's view that genetically based interventions in the unborn, even when they do not threaten life, nevertheless abrogate the natural basis of human autonomy.

The turn to Aristotle on the part of the Left as 'curious' because it flies in the face of all modern biological science, which from Lamarck onward has increasingly stressed the conventionality of species categories and the plasticity of organisms. Of course, it does not follow that we can now reverse-engineer any species as we please. Nevertheless, as the 'synthetic biology' movement suggests, scientists are spontaneously taking us along that trajectory, which draws us still further away from Aristotle's metaphysical starting point (Church and Regis 2012). More importantly, a hallmark of the modern Left's political commitment to large-scale, long-term social transformation has been a belief that human beings can be substantially changed for the better under the right circumstances. Indeed, this belief has been common to all the social sciences in the modern era from psychology and economics to sociology and political science (Fuller 2006). It is a proactionary presumption that appears to have been abandoned by the precautionary strand in contemporary Leftist thought.

If precautionaries would have us minimize risk-taking, proactionaries define the human condition in terms of its capacity to take, survive and thrive on risk, based on some calculation of benefit to cost. They argue that the value of an object or practice cannot be properly conceptualized – let alone evaluated as being 'over' or 'under' estimated – unless it has been assigned an exchange value (or price) within a particular moral economy, fluctuations within which may be reasonably seen as market-like. Indeed, it is not clear how Marxists would have been able to tell whether workers were being 'exploited' had they not operated with a sense of a 'fair wage' that could be specified in monetary terms, which in turn implies that the value of human labour is neither indeterminate nor infinite (Newey 2012). In this respect, proactionaries return to the philosophical backdrop that originally united the 'liberal' and 'socialist' branches of the Left.

Until Karl Polanyi (1944) began to seed what is nowadays the 'eco-socialist' critique of the Enclosure Acts that Parliament passed in the eighteenth century, effectively privatizing much of the British countryside, the Acts had been seen as a relatively successful albeit risky venture to increase land productivity by legally assigning personal responsibility for its maintenance, a precondition for the innovative uses and transfers of property that characterized the Industrial Revolution (McCloskey 1975). To be sure, liberals and socialists differed substantially over the impact of this development on social relations. By the mid-nineteenth century socialists had called for a 're-collectivization' of the means of production, given that private ownership had begun to settle into new class-based hierarchies just as pernicious as the old aristocratic ones that the bourgeoisie claimed to have overturned. This in turn provided the basis for the various worldwide self-styled 'Communist' revolutions of the twentieth century. However, these revolts retained the proactionary impulse. Lenin did not revert to a Rousseauian sense of 'commons' that had pre-existed private property; on the contrary, he amalgamated privately owned land into artificial persons called 'collectives' that functioned largely as the individual owners had, while taking advantage of a perceived economy of scale and a rationalized division of labour, both designed to increase productivity while short-circuiting narrow pursuits of self-interest (Scott 1998: chap. 5). In Chapter 4, our proposal of 'hedgenetics' translates this impulse into the legal framework of liberal democracy. The key intermediate move, discussed in Chapter 3, is to conceptualize our genetic material as property that one is entitled, and perhaps even obliged, to dispose of as inherited capital.

The liberal pursuit of the proactionary principle in the twentieth century was most evident in the radical doctrines of 'risk, uncertainty and profit' propounded by Frank Knight (1921), the intellectual founder of what is now called the 'Chicago School of Economics'. Today the Chicago School tends to be understood in terms of what it became in the second half of the twentieth century, in light of the influence of Friedrich Hayek and Milton Friedman, namely, an unqualified upholder of property rights in a de-regulated market environment (Davies 2010). Because Knight's original work was done before those political doctrines were set in stone, it provides an opportunity for considering a world-view very close to that of the radical Scotist interpretation of what is possible. In particular, Knight viewed the economy from the standpoint of the entrepreneur, the person who converted the 'unknown unknown' into 'known unknown' – that is, 'uncertainty' into 'risk', in the technical senses of these terms for which Knight is normally credited. Still more plainly, the entrepreneur is someone keen on marketing a product that not only attracts buyers but also sets a new standard for demand, much as Henry Ford's automobile had done for personal transport in Knight's day. However, the entrepreneur does not know how much to invest to bring about the desired result (or even whether any amount will be enough) – yet he must invest something. Whether that investment counts as 'profit' or 'loss' will be known only after the fact, and hence cannot be properly costed in advance: if you will have spent too much to achieve your goal you will receive a profit, too little a loss.

Indeed, this was why the late nineteenth-century Austrian finance minister Eugen Böhm-Bawerk (1959) had argued *contra* Marx's theory of 'surplus value' that the entrepreneur is entitled to retain all of his profits and not redistribute them to his workers. After all, the workers would have been paid their wages even if what they produced had not cleared the market. In effect, the workers' fortunes had been protected all along in a way that the entrepreneur's own were not. In that respect, the employment of labour is a necessarily non-innovative feature of entrepreneurship – that is, unless labour is subject *à la* David Ricardo to organizational efficiency savings, as in the case of Ford's assembly-line style of manufacture. The Scotist logic here is that if costs are calculable prior to investment, then you are doing no more than projecting the past into the future rather than tapping into a potential that has yet to be realized. Moreover, the learning that results from entrepreneurship, both failed and successful, tends precisely in that direction, such that uncertainty is converted into risk, luring the entrepreneur to morph into a manager of costs and benefits.

Thus, the entrepreneurial spirit always needs to renew itself by colonizing new spheres of uncertainty. By the last quarter of the twentieth century this restless spirit had acquired a name: 'venture capitalism'. But as Schumpeter (1942) predicted long before, such renewal of the capitalist spirit fuels recurrent bubbles of speculative investment, the de-stabilizing effects of which will eventuate in a precautionary social welfare state. This turned out to be the Cold War settlement designed to divest both capitalism and socialism of their most proactionary tendencies. The deep imprint that this settlement has left on philosophical intuitions over the past half-century should not be underestimated. In particular, the preeminence given to John Rawls' (1971) *A Theory of Justice* rests on his readers sharing the intuition that, in a state of ignorance (or uncertainty) about one's exact position in society, it is better for the society to be organized so as to minimize worst possible life outcomes – in effect, a precautionary welfare state. In contrast, as we hope to make clear in this book, our own political end-state is a *proactionary welfare state* (cf. Pedersen 2013).

Setting aside whether Schumpeter's prognosis was either warranted or vindicated, it is clear that entrepreneurs treat their speculative investments as a material extension of hypothesis testing, in which discovering the limits of the existing market for a line of products resembles discovering the limits of the dominant theory for a given domain of reality. The organization of labour and capital to produce an innovative product is thus akin to the construction of what Popper, after Francis Bacon, called a 'crucial experiment'. It means that a successful innovator is like a successful gambler who enters the casino assuming that he will lose – but instructively (Evans 2012: chap. 8). In both cases, psychologically speaking, one is prepared to persevere in the face of adversity, treating all setbacks as 'providential', in the language of natural theology – that is, put there for a (divine) reason. In the case of science and technology, the players have usually had sufficiently big egos and/or deep pockets or, starting in the nineteenth century, institutional protection by virtue of holding tenured academic posts to sustain a pummelling. But the motto in all these cases is the same: 'No pain, no gain'.

However, as the provocative self-styled 'black swan detector' Nicholas Taleb (2012) would put it, well-bounded disciplines (*à la* Kuhnian paradigms) and well-managed firms can easily turn 'fragile' in a technical sense: their reliability depends on the extent to which they control their environments and hence focus on 'known unknowns' but not 'unknown unknowns'. The 'known unknowns' are captured by the classic laboratory set-up, in which variables are clearly identified and experimental outcomes are explicable in terms of them, albeit in ways that may falsify the hypotheses under consideration. But then as science and technology studies researchers after Latour (1987) have shown, extending that level of control beyond the walls of the laboratory proves to be an ongoing struggle, the solution to which has been to remake the world to reflect disciplinary standards. Whatever else, the result is an incredibly inefficient way of living. It effectively converts what Kuhn (1970) called 'anomalies' into enemies of some presumptive conception of progress.

This issue is usefully put in an anthropological context by appealing to a contrast in styles of artefact construction that has been observed by practitioners of 'cognitive archaeology' (de Beaune et al. 2009). This field may be the closest thing to a science of 'intelligent design' currently on offer. From its perspective, academic disciplines – Taleb's paradigm case of 'fragile' inquiry – are instances of 'reliable systems' whose components are 'overdesigned and understressed'. In other words, the system users (i.e. academics) exert sufficient control over their environments to build in relevant responses to anticipated contingencies. In contrast, the sort of 'antifragile' inquiry championed by Taleb and other proactionaries leans towards 'maintainable systems', whose construction is sufficiently simple and nimble that their users can repair as they go along. And even when beyond full repair, such maintainable systems provide sufficient utility to enable their users to survive until a better version comes along. In the original cognitive archaeology of artefacts, the difference between, say, a 'reliable' and a 'maintainable' hunting device turned on whether the prey is targeted and timed or scattered but ubiquitous (Bleed 1986).

You cannot learn from your difficulties and errors if you insist on treating whatever resists your will in negative terms, such that you end up hiding your errors in shame. *Rather you need to see errors as investments in a better future.* While you may not be the prime beneficiary of the lessons learned from your errors, at the same time you should not suffer undue moral or legal stigma for having committed the errors. This is the spirit in which proactionaries treat the market (in the case of capitalism) and the state (in the case of socialism) as a scientific testing ground, This *modus operandi* is completely alien to the precautionary approach, whose own equally powerful appeal to science involves underscoring existing uncertainties, not with an eye to resolving them through some experimental interventions but on the contrary, to curbing the pace and scale of technological innovation itself. Although precautionaries style themselves as 'guardians of the future' (e.g. Read 2012), their tendency to use science in such an overwhelmingly reactive and negative capacity, ignore several factors that together conspire to make for what proactionaries would regard as a 'perfect storm' for future generations:

1. increasing scientific knowledge about our material constitution;
2. weakening state power over the welfare of individuals nominally under its control;
3. increasing willingness of corporate power to pick up the slack of the state's retreat, extending to the production and distribution of scientific knowledge of ourselves;
4. the lessening of corporate accountability vis-à-vis (3) as the state outsources its own functions;
5. human adaptability as a species to outcomes, such that if we do not take deliberate action, we might well sleepwalk into a suboptimal future.

As the above 'perfect storm' scenario suggests, the main obstacle facing the enforcement of the proactionary principle in a fair and equitable manner comes from increasing corporate control over the scientific understanding of humanity – including of our genetic make-up – in the form of privately owned intellectual property. Our concern here is limited to questions of the ownership and disposition of this intellectual property, not to the idea of intellectual property itself. There is no doubt that the scale and scope of 'big business' has contributed significantly, especially in the twentieth century, to fuelling scientific ambitions and human aspirations, often in the face of active resistance from academia. And no doubt much of the resulting research – ranging from molecular biology to organizational sociology – has advanced the public good. (In Chapter 3, the Rockefeller Foundation is considered in this partly redeeming light.) The problem is that it has done so only as a by-product of profit-making, which in a relatively de-regulated knowledge economy (i.e. where the state does not play an active role in overseeing intellectual property transactions) may eventuate in corporate ownership of even human reproductive capacities. This dystopic scenario was vividly portrayed in *Next*, the last novel that the best-selling author Michael Crichton (2006) published before his death. In the novel's postscript, Crichton called for the state to enact legislation that would conserve the human gene pool by outlawing its private corporate control.

Crichton, a libertarian, cast this proposal in terms of the protection of individual freedom. However, the proactionary principle, while sharing many libertarian ideas (and followers), takes the protection of individual freedom not as an end in itself but a means for the cultivation of 'humanity', understood as a being whose nature is both self- and world-transforming. (This, as we have seen, is in strong contrast to the supporters of the precautionary principle, who presume that 'Nature' sets a non-negotiable norm to which we and other living beings must ultimately conform.) The political economy required for this 'cultivation' is an entirely revamped conception of the welfare state. Instead of the historic welfare state strategy of simply discouraging risk-taking (e.g. by promoting 'healthy living'), this new proactionary welfare state would provide a relatively secure bio-social environment for the taking of calculated life risks in return for reward, repair or compensation at the personal level – as well as providing a rich data base from which society may benefit as the progress of science is expedited (Fuller 2012: chap. 2).

The securitized encouragement of these life risks can be justified in proactionary terms as extending the duties of citizenship to include participation in 'scientific research', now understood as licensed both to research facilities (e.g. laboratories) and individuals (i.e. self-experimenters). This argument is already being made by bioethicists sympathetic to transhumanism (Chan et al. 2011), and will be further pursued in the final chapter of this book. Two precedents from the history of democratic politics stand out here: (1) the duty of national service as a concomitant of the right to participate in political life (cf. you have a say about the future of scientific research, especially as it bears on humanity's self-transformation, by virtue of your having acquired a stake in it); (2) the enforcement of literacy as a capacity required to exercise both the fundamental human right to self-expression and the state obligation of public accountability (cf. the ongoing recording of the consequences and responses to the risks one undertakes).

4 Conclusion: Marking the rotation of the ideological axis

The main bone of contention between Left and Right in the modern era has been the state's prerogative to deliver *social justice*. On the whole, this is a duty that the Left recognizes but the Right does not. Even so, the reasons for taking either position have varied significantly within each camp. The technocratic Left has seen social justice as part of a larger agenda of social progress, whereas the communitarian Left has tended to focus on the need to secure each person a decent quality of life. The libertarian Right has dismissed the very idea of social justice as inherently authoritarian, while the conservative Right has sought a more 'natural' form of justice in such pre-modern institutions as the church and the family.

In the old Left–Right dialectic, the 'state' stands for a peculiarly modern corporate agent that has been granted perpetual authority to make laws, administer to basic needs, and promote prosperity for a human population living within a generally recognized territory. To older readers, this point will seem obvious. However, in a world where people are increasingly affiliating across national borders and species boundaries, and where some even aspire to discover extraterrestrial intelligence, the state might seem much less salient, even an outright obstacle. From this perspective, the basic dilemma that has defined the Left–Right divide for the past two centuries – *Shall we extend or limit state power?* – starts to look a bit stale.

On top of this, there is the state's striking failure to deliver on its own promises of social justice – what Marxists, in a great piece of rhetorical distancing, have dubbed 'the fiscal crisis of the state' (O'Connor 1973). The persistence (if not resurgence) of poverty, inequality and ethnic discrimination, despite a succession of well-intentioned, often well-financed and indeed sometimes partly successful welfare programmes, has put the Left on the defensive since the 1980s. The situation here recalls one that the precursors of the modern Right faced in the 1780s, when the long-term viability of the Church-backed monarchies of Europe came into serious question. That loss of faith resulted in the French Revolution and the advent of the Left–Right divide. The 1980s were likewise marked by widespread scepticism of the dominant narrative – no longer the providential hand of God but rather the march of human progress. And, in both cases, the leading political response was a paternalistic attitude towards the poor that focused on securing their material existence in what was increasingly presented as a high-risk environment, rather than aiding their passage up any spiritual or social ladder. Polanyi (1944) provides a rich sense of the affinities between classic 'Tory' paternalism and the modern welfare concerns of the communitarian Left. These tendencies are due for a formal reunion within the Green movement, the first steps towards which have already been made by sophisticated conservative thinkers (Scruton 2012; cf. Turner 2010).

Indeed, Karl Polanyi, the Austrian political economist who is now seen as a founder of modern economic anthropology, is reasonably regarded as a founder of this 'precautionary socialism', since he grounded socialism's redistributivist ethic less in abstract considerations of universal justice or even allocative efficiency than its historically normal (what a conservative would call 'traditional') character, the violation of which by *both* the modern state and the modern market is then invoked to explain the striking resource inequalities that exist in today's societies. Moreover, there is even a self-described 'liberal' side to the emerging precautionary ideology, which can only be glanced here. This species of liberalism derives from the failed 1848 European revolutions, becoming pronounced in the post-1918 version of the Austrian School of Economics (i.e. Mises, Hayek, Schutz, etc.), one that is profoundly sceptical of the human capacity to control, or even quantify, large-scale social processes, which renders meaningless any sense of 'collective learning' above and beyond the social arrangements that manage to survive in the course of time. Such liberalism, while 'libertarian' in name is 'reactionary' in effect (Hirschman 1991).

In terms of the ideological colour scheme, we may be seeing here a *Green versus Black* ideological divide – for precautionary and proactionary, respectively – replacing the *Blue* (Right) *versus Red* (Left) instituted in the aftermath of the French Revolution. (Unfortunately, for younger Americans, this chromatic clarity is somewhat obscured by the 'Red State–Blue State' distinction that the respected liberal television journalist Tim Russert introduced in the 2000 US presidential election to shift the significance of the colour red from Communist to Redneck.) The colour scheme of the European spectrum was originally divided between blue and red because blue blood was seen as 'purer' (i.e. deoxygenated) as it flowed into the heart via the veins, whereas red blood reflected its contamination by the air as it flowed outward from the arteries. Needless to say, this was a late eighteenth-century way of seeing matters. So blue blood was valued for recycling the past without being corrupted by the present. We now see the normative implications of these physiological facts in the exact opposite way, and the ambiguities produced by that transition in sensibility has no doubt helped to stabilize the blue–red spectrum, enabling each side to spin the metaphor to their own advantage, as in the association of red blood with 'fresh blood'.

The colours most often invoked for the emerging precautionary and proactionary ideologies, *green* and *black*, are meant to stand, respectively, for the Earth and the heavens. While the former may be self-explanatory given the precautionary principle's historic association with the ecology movement, the latter may require further explanation. The phrase 'Black Sky thinking' was coined in a 2004 study by the centre-left UK think-tank Demos (Wilsdon and Mean 2004). Over the last decade the phrase has been increasingly used to capture schemes to make the entire universe fit for human habitation. Had this proposal been made a generation earlier, it would have been understood in a rather abstract way: namely, our power to access anything from any place at any time, the sort of ubiquity that smart-phone owners nowadays routinely expect. The focus then was on communication not transportation. Nowadays, however, self-described Black Sky thinkers at the US-based Lifeboat Foundation and Icarus Interstellar aim for something considerably more flesh-and-blood: to incorporate aspects of Earth's natural habitat in vessels capable of an indefinite journey through space. This goal is almost the exact opposite of the Greens, who would fit human aspirations to what nature in its earthbound form can be reasonably expected to sustain in the future.

In that case, where should we look to see the libertarian Right and the technocratic Left making common cause in a Black-sided world? A good place to start is the 'DIY Bio' (Do-It-Yourself Biology) movement that tries to reverse-engineer organisms in the same spirit as one might try to convert an ordinary car into a drag racer. George Church, the Harvard medical geneticist who promotes the resurrection of extinct species, sees in DIY Bio the first truly grassroots advanced science, 'synthetic biology' (Church and Regis 2012). DIY Bio enthusiasts happily transfer already existing DIY 'open source' attitudes from computer code to genetic code, which renders the organism a site of tinkering and troubleshooting. Their default normative stance is libertarian, evading and sometimes even mocking the scientific short-sightedness of conventional morality, not least of academic 'institutional review boards' that oversee research ethics.

But unlike right-wing libertarians, synthetic biologists also generally believe in grand 'technological fixes' for solving the world's problems, the most headline-grabbing of which has perhaps come from Church's lifelong rival, Craig Venter, who has proposed solving the world's food problems through hydroponics (i.e. soilless agriculture). Such feats of what Silicon Valley denizens call 'visioneering' (Corbyn 2013) are regularly derided by the journalist Evgeny Morozov (2013) as 'solutionism'. In whichever guise, regularly featured in TED talks, Black-siders can easily come across as unholy hybrids of Friedrich Hayek's radical liberalism and Saint-Simon's utopian socialism, the existence of which would cause the head of the author of Hayek (1952) to explode. Without denying (or, for that matter, decrying) that characterization, there are more politically palatable expressions of the same attitude. Consider the Breakthrough Institute, a US-based environmental think-tank that sees capitalist investment in advanced science and technology as the key to humanity's survival on Earth, and possibly beyond. Most of its members appear to be Left-leaning, but whatever commitment to socialism they might have, it is one that sees itself as building upon – not abandoning – capitalism, very much as Saint-Simon and, yes, Karl Marx once did. Their various cases for experimental approaches to new energy sources reveal a fundamental optimism about the human condition, whereby every new existential threat is an opportunity in the making.

We have seen that proactionaries would re-invent the welfare state as a vehicle for fostering securitized risk-taking, while precautionaries would aim to protect the planet at levels of security well beyond what the classic welfare state could realistically provide for human beings, let alone the natural environment. Taken together, these two opposing innovations to the concept of welfare imply a rejection of the classic welfare state ideal that humans might procreate at will in a world where their offspring are assured a healthy and safe existence. For all their substantial disagreements, both poles of the emerging ideological order dismiss this prospect as a twentieth-century fantasy that was only temporarily realized in Northern Europe for a few decades after the Second World War. Not surprisingly, conventional political and business leaders are not entirely comfortable with either the precautionary or the proactionary principle, which in turn helps to explain their lingering attachment to some version of the old ideological Right–Left divide. After all, precautionary policymakers would have business value conservation over growth, while proactionary policymakers would have the state encourage people to transcend current norms rather than adhere to them. A precautionary firm would look like a miniature version of today's regulatory state, whereas a proactionary state would operate like a venture capitalist writ large.

The classic welfare state's loss of political salience reflects a massive transformation in humanity's self-understanding, albeit in two diametrically opposed directions. Together they constitute the self-divided entity that Fuller (2011a) has dubbed 'Humanity 2.0'. Both sides in this self-division pull away from 'Humanity 1.0', the entity enshrined in, say, the United Nations Universal Declaration on Human Rights. One side pulls 'genealogically' to extend similar rights, concerns, etc. to those with a common evolutionary past; the other side 'teleologically', to extend similar rights, concerns, etc. to those with a common progressive future. The former, biased towards biology, goes with the precautionary principle; the latter, biased towards technology, goes with the proactionary principle. Precautionaries aspire to a 'sustainable' humanity, which invariably means bringing fewer of us into existence, with each of us making less of an impact on the planet. In contrast, proactionaries are happy to increase the planet's human population indefinitely as nothing more or less than a series of experiments in living, regardless of outcomes. Whereas precautionaries would reacquaint us with our humble animal origins, from which we have strayed for much too long, proactionaries would expedite our departure from our evolutionary past – in some versions, the Earth itself, if we succeed in colonizing other planets. At the very least, proactionaries would re-engineer our biology, if not replace it altogether with some intellectually superior and more durable substratum.

The precautionary–proactionary divide has the potential to shift the ideological axis by 90 degrees. The Right is currently divided into traditionalists and libertarians; the Left into communitarians and technocrats. In the future, we suggest, the traditionalists and the communitarians will form the precautionary pole of the political spectrum, while the libertarians and technocrats the proactionary pole. Scruton (2012) and Kelly (2011), respectively, are early manifestations of these new configurations. These will be the next Right and the next Left – or, rather, the Down and Up. One group will be grounded in the Earth, while the other looks towards the heavens. An inspiration for this shift in imagery in ideological 'wings' is E.M. Esfiandiary (1973), the Iranian futurist who portrayed the people we call 'proactionaries' as 'up-wingers' and 'precautionaries' as 'down-wingers'. The implied rotation in the ideological axis is depicted in Table 1.

Table 1 Reorienting the wings of the ideological axis

	LEFT WING	RIGHT WING
UP WING	Technocrat	Libertarian
DOWN WING	Communitarian	Traditionalist

2

Proactionary Theology: Discovering the Art of God-Playing

1 Introduction: The biblical roots of playing God

Just as a religious temperament does not demand a belief in a supernatural deity, a belief in such a deity does not require a religious temperament (cf. Nagel 2010: chap.1). Spinoza's notorious philosophical radicalism, now routinely seen as a wellspring for the eighteenth-century Enlightenment, capitalizes on just this disjunction. Spinoza himself sided with those who found in nature a sufficient source of religious devotion without having to invoke a higher intelligence that underwrites it all. For him, God is simply coextensive with nature. This was the original meaning of 'naturalism', which with the help of Ockham's Razor was then used to launch modern atheism in the nineteenth century, as people began to think: 'Wouldn't "Nature", understood in its totality, suffice as the name of God?' The authors of this book, on the other hand, stand with those who locate the 'best explanation' for nature in the workings of the sort of anthropocentric yet transcendent deity favoured by the Abrahamic religions (Meyer 2009; Fuller 2010).

These religions – Judaism, Christianity and Islam – are often called 'religions of the book', and for very good reason (Fuller 2006: chap. 11). Their literal-mindedness reflects a preoccupation with the terms of exchange between the deity and its privileged offspring, the human. The histories of these religions are thus punctuated by talk of covenants, contracts and codes – as well as the presence of great lawyers – from the great Talmudic scholars and the Muslim heretic Averroes to the modern reformed Christians, John Calvin and

Francis Bacon. As the scope and power of digitization extends over more of reality, the classic distinction between the organism and its blueprint is being blurred, perhaps most dramatically in the prospect of 3-D printing technology enabling us to project novel biochemical complexes into being, a kind of Holy Grail of synthetic biology, whereby we would finally – and literally – acquire the divine capacity to create through 'The Word'.

We believe that the original motivation for the West's Scientific Revolution – the radical version of Christian self-empowerment championed by the Protestant Reformation – remains the best starting point for motivating the contemporary transhumanist project. We take the scientific establishment today to function as the Roman Catholic Church did in the sixteenth and seventeenth centuries (cf. Fuller 2010: chap. 4). Both grant only a qualified licence to the extension of otherwise agreed doctrines. In both cases, the orthodoxy's need to protect its epistemic authority has been the overriding consideration. Against this backdrop the transhumanists are among the new Protestants, pushing the established doctrines harder and farther. Without denying the general point that ideas have flourished outside the context of their origins, nevertheless success requires the idea to be sufficiently motivated to persist even as its empirical track record remains chequered. This is the spirit in which the search for intellectual origins should be understood. More than antiquarian – or, for that matter, critical – interest, origins provide vital information about sustaining motives that, in the case under consideration, attempt to shift the burden of proof to those who believe that the apparent rationality of the universe is *not* the product of some sort of human-friendly intelligent design. But what else could possibly justify transhumanism other than a literal belief in our own capacities for apotheosis?

After all, much simpler means are already at our disposal (at least in terms of the redistribution of income and resources) to alleviate human poverty and produce a more ecologically sustainable world. Moreover, the empirical track record of the vanguard forms of twentieth-century science on the back of which transhumanism is promoted leaves something to be desired: every new breakthrough in genetics, nuclear energy and biochemistry more generally has delivered equal measures of harm and benefit, albeit often to distinct parts of the human community. Indeed, no one seriously denies that the existential risks facing the survival of our species are science-based, albeit more by effect than by design. But equally there would not be so many members of *Homo sapiens* placed at risk, had it not been for this dogged pursuit of science. In other words, we seem to be the ultimate cosmic gamblers, perhaps with the deepest pockets, who do not know when to quit. Instead, we somehow manage to ride out every boom and bust in our fortunes.

But none of this is yet to commit to more specific religious doctrines about the *modus operandi* of this legalistic deity, or for that matter the spirit in which humans should engage with this deity. (For a survey of options, see Brague 2007.) Moreover, as we shall shortly see, the relevant theologies that might inform such a commitment are bound to challenge religious orthodoxies. Nevertheless, we aim to reverse the prejudicial positioning of the 'supernatural' as some irrationalist misunderstanding of the 'natural'. On the contrary, we argue that science and humanity more generally have been promoted by presuming that the world is intelligently rather than unintelligently designed, which is our gloss on the allegedly blind 'design without a designer' processes of natural selection, which for the past half-century have been the orthodoxy in popularizations of biology (e.g. Dawkins 1986).

Theomimesis ('God-playing' in Greek) is our neologism for attempts to acquire God's point-of-view, which is to say, literally 'getting into the mind of God' and thereby be capable of 'playing God', the former phrase still resonant in physics and the latter in biology. (The concept of theomimesis, introduced in Fuller 2008b, was expanded in Fuller 2010, 2011a, 2012.) Indeed, these scientific descendants of theomimesis track what might be called the *transhuman telos*, namely, the tendency to improve upon, if not perfect, virtues incipient in the human condition (Passmore 1970). However, in what follows, we focus on the theological dimension of the issue, in which those 'incipient virtues' refer to our divine potential. The deity in question is Abrahamic, indeed, the 'monotheistic' deity that eighteenth-century Enlightenment thinkers such as John Toland in Britain and Gotthold Lessing in Germany abstracted as the rational common core of Judaism, Christianity and Islam. In practice, this monotheism was usually but not always heavily biased towards some version of the Christian deity (cf. Elmasarfy 2009). This deity is distinctive in that it transcends the world it has created yet its nature is sufficiently close to our own that we might reasonably aspire to approximate the deity's virtues.

Theomimesis requires a *transfiguration* of the human condition. 'Transfiguration' is a term from Christian theology referring to the moment when Jesus discovered that he was divine, an episode that the Gospels recount in terms of his ascent up a mountain (Matthew 17:1–9, Mark 9:2–8, Luke 9:28–36). Christianity's distinctiveness in the Abrahamic religions lies in its specification of an individual who is at once human and divine. In Judaism, Moses is portrayed as clearly separate from God but trying, however imperfectly, to follow God's dictates. To secular eyes, this looks like an internal psychic struggle, in which the reality principle (God) eventually triumphs. In Islam, on the contrary, Muhammad is portrayed as someone who is overtaken by God, undergoing a complete personality transformation that renders him a divine vehicle. (The Arabic word 'Islam' means submission.) However, interestingly, Jesus is 'transfigured' by his re-conceptualizing from first principles what he had been doing all along. The Transfiguration episode occurs not at the start but in the midst of Jesus's ministry. The experience amounts to a religious version of the 'Gestalt switch' that Thomas Kuhn (1970) popularized with regard to what happened when, say, scientists started to 'see' the Earth not as the centre of the universe but as just another planet circling the sun. In other words, Jesus was not told that what he had been doing was wrong; rather he was compelled to re-think what he had been doing in terms of what he should do next (cf. the 'new riddle of induction', Fuller 2012: chap. 1).

That Kuhn's theory of scientific change should provide a model for understanding the Transfiguration of Jesus is unsurprising, if we consider that, again in contrast to Moses and Muhammad, Jesus grounded his ministry in explicit dissent from rabbinical orthodoxy, with the aim of recovering the lost spirit of a shared religious tradition. In other words, a 'paradigm' (Judaism) had been in place that, despite its longevity, was no longer meeting people's spiritual and material needs. Here it is worth recalling that, for Kuhn, a 'paradigm shift' is about 'going back to basics' – but *not* in the sense of returning to some brute facts that had been obscured. Rather, it is about re-motivating inquiry by seeing all that had been known in a new light that expands one's epistemic horizons; hence, the etymological kinship of *theosis* (i.e. the Greek term for Jesus's realization of his divinity) and 'theory' and even 'theatre': the implication in all these cases is that one adopts a position that transcends one's immediate condition. Call it, if you will, 'out of body' or the 'view from nowhere'. Philosophers, psychologists and, most recently, neuroscientists have sought the gland, gene, or circuitry capable of triggering such a superior mental state in any of us (Fuller 2014). However, the quarry is less the feeling of transcendence itself than the image of the world that appears when one is in such a state. For example, when Isaac Newton called space and time God's 'sensorium', he envisaged a standpoint from which every moment – be it past or future, near or far – is equally present to the viewer. That standpoint would be the proverbial 'Mind of God'.

2 Theomimesis in the modern sacred and secular imaginations

In the modern period there have been two largely secular ways of engaging in theomimesis. They derive from the two most enduring Christian heresies, against which Augustine of Hippo originally defined (often very subtly) Christian orthodoxy in the fourth century AD: *Pelagianism* and *Arianism*. In both heresies, Jesus is presented as a role model available to everyone without the need for prior clerical approval. What divides the two heresies is that the former offers a materialistic strategy to raise ourselves and our world up to suitably divine proportions, while the latter proposes a more idealistic strategy of divesting ourselves and our world of that which does not promote the achievement of divinity. The first heresy is due to the Celtic lawyer Pelagius, who argued that all humans possess the divine potential to rise from their fallen state and build a Heaven on Earth. The second heresy is due the Libyan bishop Arius, who denied that Jesus had a special relationship with God that other humans lacked. Rather, Jesus simply discovered how all of us might come to recover our lost unity with God.

With hindsight we might say that the Pelagian heresy cashed out Christian salvation as technological progress, and the Arian heresy (of which Newton was explicitly accused) as scientific progress. The Pelagian heresy begins with the disciplining of nature, including the breeding of creatures for human use, and moves to a more systemic transformation – some would say destabilization – of the physical environment so as to allow for the coexistence of increasing numbers of humans, the final frontier being the radical enhancement of our own physical and mental powers. This is a trajectory common to the more optimistic Enlightenment *philosophes*, the great industrialists and latter-day transhumanists (Noble 1997). For its part, the Arian heresy begins with the disciplining of the mind, abstracting away from our diverse sensory modalities and focusing Plato-style on that which we can be known by a being unencumbered by an animal body. Thus, instead of trying to *humanize nature*, *à la* Pelagius, the Arians wanted to *denaturalize humanity*. The former is epitomized by 'biomimetic' projects that enable the extension of human powers (including longevity) by incorporating elements of the rest of nature (Benyus 1997). The latter is captured by what transhumanists have called 'morphological freedom' – that is, the prospect of the human essence migrating across material forms, including from carbon to silicon incarnations and possibly pervading the entire cosmos, *à la* Kurzweil (2005). Such a scenario suggests more an 'Earth in Heaven' than a 'Heaven on Earth'.

Perhaps the first modern secularization of the theomimetic moment came with René Descartes' *cogito ergo sum*, which can be read as re-enacting how God's own existence is demonstrated by calling the world into being. This capacity for divine re-enactment was formalized by late seventeenth-century 'rationalist' philosophers in the Cartesian tradition, most notably Nicolas Malebranche and Gottfried von Leibniz. Malebranche in particular spoke of our 'vision in God', the features of human cognition (so-called *a priori* knowledge) that by virtue of overlapping with the divine intellect, enable us to access truths (e.g. of mathematics) that we cannot access through sensory experience alone. Kant put this theologically challenging idea in more diplomatic terms by speaking of God as a 'regulative ideal of reason'. As we saw in Chapter 1, more plain-speaking theologians and philosophers have followed the great fourteenth-century Franciscan theologian John Duns Scotus' theory of 'univocal predication' whereby divine attributes differ from human ones only by degree not kind. Thus, God may be 'all knowing' but the sense in which God 'knows' is the same as our own, except we 'know' to a much lesser degree. This in turn allowed for direct comparisons between human and divine conditions of being, and hence a trajectory of progress, typically presented as a project of species self-improvement from Adam's Fall to (possibly) the construction of Heaven on Earth (Funkenstein 1986: chap. 2).

Two versions of humanity's journey towards self-improvement have enjoyed considerable secular influence in the modern era: (1) Leibniz's theodicy, which would understand Creation in terms of a divine utility function that tolerates many local harms in service of ours being 'the best possible world'; (2) Hegel's philosophy of history, which makes temporality constitutive of God's own self-realization, which means that Creation itself is incomplete as long as the distinction between God and humanity remains. Historically these two secular theomimetic projects are known as (1) *capitalism* and (2) *socialism*. The theomimetic agents are correspondingly known as (1) the *entrepreneur* and (2) the *vanguard*. 'Moral entrepreneurship' (the recycling of evil acts into greater goods) captures the ethical horizons of such agents (Fuller 2011a: chap. 5; Fuller 2012: chap. 4).

But once religious believers started to take seriously that the actual world might reflect the design of a divine intelligence that is literally 'superhuman' (i.e. the ultimate extension of human intelligence), disaffection quickly set in, as God seemed to operate on the principle that the end justifies the means. For humans this normally means 'unscrupulous'. To be sure, the Jesuits had already seen this problem in the Counter-Reformation, proposing the 'doctrine of double effect', which aims to dissociate what one *intends* and what one *anticipates*. Thus, God always intends good, and through his all-powerful nature can bring about good, but the good of primary interest to the deity is ultimate good, not immediate or transient good. These lesser goods are related directly to matter, which for God is always a negotiable instrument. So while it is unfortunate that many must suffer and die, this would happen in service of an end to which they themselves would have agreed (had they been asked). Both military invasion and land dispossession have been justified on these terms.

Thus, it is perhaps unsurprising that Charles Darwin renounced his lingering attachment to Christianity once he studied closely William Paley's *Natural Theology*, which openly endorsed Reverend Thomas Malthus's views of divine population control through resource constraint (Fuller 2011a: 175). Of course, Darwin retained the substance of Malthus's theory – now re-branded as 'natural selection' – but he refused to believe in the existence of a deity who would allow so many members of so many species to endure such miserable existences. Darwin's morally fuelled atheism thus led him to a kind of Neo-Epicureanism that dissipated divine responsibility in a sophisticated version of metaphysical indeterminacy.

However, the decline of theomimesis among professional theologians is trickier to explain. Most embraced Darwin's empirical findings and hypotheses without losing their faith. Indeed, the strong pro-science orientation of theologians in the half-century following the publication of *On the Origin of Species* is exemplified in Adolf von Harnack, who served as political midwife to the birth of the Kaiser Wilhelm (now Max Planck) Institutes, the pioneer vehicle for state–industry–academia research collaborations. Rather like his contemporary Ernst Mach vis-à-vis the Newtonian orthodoxy in physics, Harnack championed an empirically informed, relentlessly critical approach to Christian dogma that by removing all the errors that have accrued through history would eventually hit upon God's original message (Gregory 1992).

But this close bond between science and theology was undone by the German scientific community's strong presence in the First World War. The unprecedented devastation of the 'Great War' led many theologians to doubt that science could be treated as the pursuit of God by critically informed and technologically enhanced means. From this conclusion arose the sort of 'fideism' championed by one of Harnack's students, Karl Barth, which in the name of humanity's fallen state rendered theology a self-referential discourse without any aspirations or accountability vis-à-vis the world as understood by science. As the Great War had shown (at least to Barth's satisfaction), the very attempt to redeem theological claims in scientific terms was itself to court evil.

Moreover, Barth would seem to have been further vindicated as the twentieth century wore on. Harnack's own brand of dissenting Christianity saw in the Marcionite heresy an early radical understanding of the religion's attempt to divest itself of all parochialism – including its own Jewish origins – in its quest to be the faith of a universal humanity. This gradually morphed into a high-minded version of the anti-Semitism that predominated in the Nazi period. Nevertheless, without denying Barth's principled stance against the Nazis, his overall influence has been regrettable from the standpoint of an open borders policy whereby theology might frame the questions that science asks. Indeed, his inward, even reactionary, turn helps to explain the ease with which science's staunchest twentieth-century philosophical supporters, the logical positivists, could get away with asserting the 'non-cognitive' status of religious discourse – without hearing much theological complaint in return. Today's 'New Atheists' merely raise the positivist ante by querying the whole point of religion, once its cognitive aspirations have been abandoned (Dennett 2006).

In response, religious believers do themselves no favours by acting as if theology were literary criticism applied to an especially vividly experienced form of imaginative writing, otherwise known as 'fiction'. Still worse offenders are those 'religious pluralists' who declare intelligent design theory – the natural heirs to the theomimetic tradition sketched above – to be 'bad science and bad theology' (e.g. Armstrong 2009). This must count as the 'Big Lie' promulgated in the contemporary science–religion debate, as it merely serves to throw honest inquirers off a trail that would lead back to a literal construal of theology as the 'science of God', in which case one might expect theories about God's nature to have scientifically tractable consequences, just as the protagonists of Europe's seventeenth-century Scientific Revolution had thought. In this context, the physicist Frank Tipler (2007) must count as someone who continues to live up to this ideal, despite his outlier status in both science and theology. Tipler is the most dogged proponent of a metaphysically full-blown version of the 'anthropic principle', which supposes that the sheer presence of beings (i.e. humans) is a product of 'fine-tuning' whose prior probability is so small that it suggests the presence of a higher intelligence that might be grasped (Barrow and Tipler 1988).

3 The four theological principles underwriting theomimesis

Although the practices associated with 'playing God' in the modern era – typically involving power over the creation and disposal of life – appear increasingly detached from their original theological moorings, the distinctive cast of those activities are still best understood as bearing the traces of those origins. So, let us return to Augustine's interpretation of the first three chapters of Genesis. In particular, he stressed four points of lasting significance, which theomimesis has turned to great secular effect:

1. *A strong reading of our having been created 'in the image and likeness of God'.* The Jews had originally read this passage as a naturalistic explanation for why God found it so much easier talking to humans than other creatures, while Muslims took it to mean that humans were essentially God's robots, ideal vehicles for the conveyance of the divine message, as in the composition of the Qur'an by the functionally illiterate Muhammad. However, Augustine, following Paul (Romans 5), read this passage to anticipate Jesus, understood not as another Jewish prophet but the 'Son of God', which encouraged the idea that humans could be godlike. In the short term, this idea begat many of the heresies that Augustine sought to stamp out as Bishop of Hippo, but it opened the door to query what it might mean to live *in imago dei*: are we to live our lives according to the dictates of Jesus's anointed successors, or are we to rediscover Jesus for ourselves in each new generation? In short: *Peter's* or *Paul's* way of spreading the Gospel (Fuller 2008a: chap. 7). The former casts divine communication as flowing through dynastic delegation, the latter through a process more akin to public broadcasting. Nowadays the choice looks like Catholicism versus Protestantism but the division was already present at the medieval founding of the universities, with the Dominican stress on natural history and the Franciscan stress on optics (Fuller 2011a: chap. 2). But equally it captures two scholarly ways of reading the Bible 'literally': as a historical document or as a theatrical script. In the one case, we *recover* the Word (i.e. we find out what the author meant back then); in the other, we *re-enact* the Word (i.e. we find out what the author would mean now). It also captures the difference in world-view between the Renaissance and the Enlightenment. By the late nineteenth century, this had crystallized as the distinction between 'idiographic' and 'nomothetic' sciences in secular academia, the one focused on the archive, the other the laboratory.
2. *A stress on the qualified nature of God's forgiveness of Adam's sin.* Augustine interpreted Adam's divine death sentence, which extended to all his descendants, as implying that his sin was forgiven yet not forgotten: each generation is born with a version of

Adam's sin that they must then somehow redeem for themselves, presumably with divine approval and preferably with the aid of Jesus. (Contrast this with the account provided in Qur'an 7: Adam's debt is cancelled in return for perpetual submission, the literal meaning of 'Islam'.) A key feature of what Augustine coined as 'Original Sin' is the specification of Adam's failure as being one of judgement, not action. Adam and Eve had erroneously trusted the serpent's plausible reasoning. Henceforth there would be grounds for questioning one's understanding of words, even though God had called things into existence. Many of the distinctive preoccupations of modern science flow from this sensibility, including these three: (a) the search for a pure language of thought that overcomes the noise of natural languages; (b) the concern with distinguishing genuine causal relations from their virtual ('evil') twin, empirical correlations; (c) the desire to mitigate our inherent imperfections by reverse engineering the bases of our material nature, tellingly called 'genetics'. Common to these preoccupations is the idea that Evil, though radically different in nature from Good, is quite similar in its appearance. Thus, considerable ambiguity has dogged the moral standing of science 'pursued for its own sake', as this most ennobling of human endeavours can easily tip over into inhumane acts, as in the Faust legend.

3. *The placement of equal emphasis on the perfection of divine creation as a whole and the radical imperfection of its parts.* The story of Creation is presented as a piecemeal process, in which God appears to weigh options, given the (undisclosed) limitations of the material medium within which creation must occur. With the rise of Islam, the divine *modus operandi* comes to be compared to a chess game in which the grandmaster always win but by means dictated by the position of the pieces on the board. In any case, the existence of such options suggests that God is an 'optimizer' or, as economists say, a 'constrained maximizer', whose ideal solution to the problem of Creation requires many calculated trade-offs that, when taken in isolation, may appear unbecoming of a supreme deity, if not downright evil. The 'will' was the name coined for a divine faculty, also present in humans, which realizes the outcomes of such calculations. But it is arguably just such knowledge that God was trying to hide from Adam when he forbade him from eating of the Tree of Knowledge of Good and Evil

because he wished humanity to be innocent of the 'dirty work' of creation – namely, the possibilities that had to be forgone to enable things to be as they are. However, once Adam had been persuaded to eat of the forbidden tree, the only possible (if at all) way back into God's good graces was by discovering – however fitfully and imperfectly it was likely to be – the divine plan to which the forbidden tree had promised access. This meant capturing in temporal terms (i.e. over many generations) an achievement that God had accomplished from a standpoint equidistant from all places and times (i.e. 'the view from nowhere'). The signature legacy of humans trying to reverse-engineer the intelligence behind divine creation is the idea that scientific inquiry should aim for the most economical set of universal laws, where such laws are understood as covering not only generalizations of experience but also counterfactual possibilities (i.e. those that God considered but rejected as well as those that God has left for humans to exploit).

4. *A conception of God's* modus operandi *as* creatio ex nihilo *('creation out of nothing'), which serves as a model for rational human agency.* Of course, humans cannot literally make something out of nothing but the divine ideal may be approximated by operating according to two distinct principles of efficiency, the former applying to humans as original creators, the latter to us as reverse-engineers of divine creation: (a) diminishing marginal cost per unit produced, such that the majority of effort occurs with the initial production of blueprints or elements that constitute everything that follows; (b) yielding greater rate of return, such that power is increased while consuming less matter and energy, or what the engineer Buckminster Fuller (1968) called 'ephemeralization', which in the recent past has been perhaps most evident in computer-based miniaturization. These two principles are interrelated in the classical Humboldtian mission of the university, in which teaching is effectively defined as the 'creative destruction' of research by distributing the benefits of research as widely and cheaply as possible (Fuller 2009: chap. 1) The original medieval universities were staffed by the Christian mendicant friars, the Dominicans and Franciscans, who were seen as God-like in their 'poverty' (*povertas*) because they made the most of whatever they were given, including at the intellectual level. In other words, they spread as widely as possible their knowledge, which may have been initially acquired by laborious

research. But notice that you can achieve more by doing less, only if you have done something – however inefficiently – in the first place. This might count as a thermodynamic definition of learning, whereby error is a precondition to progress. In this respect, the entrepreneur may be seen as the economy's answer to Maxwell's Demon who divides a current practice into a temporary loss that is incurred through speculative investment and an efficiency gain that will be increasingly realized over time.

4 Conclusion: The four styles of playing God in today's world

Classical forms of atheism – of the sort associated with the Epicureans and even David Hume – have been about lowering humanity's expectations, largely as a pain-avoidance strategy in a world where the very aspiration to divinity is seen as a potentially cruel illusion. 'Atheism' has only ever contributed to the advancement of science and the empowerment of humanity when humans have arrogated for themselves divine powers, while refusing to credit God for the inspiration; in this respect, such self-avowed 'atheist' movements as idealism, positivism, Marxism, etc. are properly called 'Promethean' (cf. Fuller 2010: chap. 6). And it is just these crypto-theistic 'atheists' who need to figure out which game they are playing when they are 'playing God'. A tableau of four possibilities that have been generated over the past 300 years are presented below and are explained in what follows. The possibilities capture the relationship between the deity and humanity, as understood along two dimensions:

(1) 'Transcendent' vs. 'Immanent' refers to whether God's nature is separable from human nature – yes and no, respectively. Thus, the 'transcendent' deity is properly described as 'supernatural' and the 'immanent' one as 'natural'. The former case implies a dualism of mind and matter as separate substances, the latter a convergence of mind and matter, whereby mind is taken to be the form of matter.
(2) 'Static' vs. 'Dynamic' refers to whether the God-human relationship changes over time – again, yes and no, respectively. Thus, a 'static' relationship can be maintained either because God stands above space and time or because God is the natural order to which all of creation ultimately returns, while a 'dynamic' relationship can be the result of some explicit transactions between God and humans

Table 2 The modes of playing God

	STATIC	DYNAMIC
TRANSCENDENT	Deism	Clientelism
IMMANENT	Ecologism	Expressivism

or that humans are the vehicle through which God fully comes into being. What follows is a motivational explanation of the four positions, each possessing its own prima facie plausibility (see Table 2).

'Deism' was the original word for 'theism', but by the late seventeenth century its referent had extended from a belief in God to a capacity to see things from God's point-of-view. After Newton's friend, John Locke, and Locke's French disciple, Voltaire, Deism came to be seen as the modern version of the Arian heresy. In this form, God appears very much as an engineer whose intelligent design is sufficiently apparent if one studies nature with due diligence that there is no need for some other pure faith-based form of inquiry. Deism has been often satirized as promoting a *deus absconditus*, that is, a deity who once he sees that nature is as he would have it, disappears because his presence is no longer required. Thus, physics makes theology redundant. Deism is open to this depiction, as the deity's construction of the best possible world involves anticipating each realizable possibility – specifically, the various trade-offs and counter-balances that would need to be managed. After all, a world that simultaneously optimizes across all the virtues cannot maximize any single one of them. It is a failing of humans to fetishize single virtues, such as wanting to be fastest runner or calculator, regardless of its effects on other equally necessary aspects of life. Yet God's optimization strategy is no mere call for moderation in all things, à la Aristotle. It is about getting the proportions right and compensating for them when they are not. One device that captured the imaginations of the early Deists was the *orrery* – that is, a mechanical model of the solar system, in which all the planets moved in proportion to their orbits when you moved any one of them. Humans are born into this self-regulating universe in the spirit of heirs capable of building their own worlds on the back of this divine capital, which lay behind the fashionable eighteenth-century drive to create a 'prosperous commonwealth'. Like a provident parent who wants to respect his offspring's autonomy but also like a careful investor safeguarding against complete disaster, the deity does not interfere with humanity's use of that inheritance. But God has rendered Creation foolproof just in case humanity fails to make the grade.

The **Clientelist** model of God-playing was dubbed the 'Limited Liability God' by the great Victorian liberal, John Stuart Mill, who appeared to half-believe in it. This is an updated secular version of the various deals ('covenants') that God appears to have struck with humanity in the Old Testament. Accordingly, God outright needs humans to complete creation because he's not a master of detail. He creates only blueprints, which correspond to natural laws. But these by themselves say little about how they might be implemented in practice. The laws of physics are compatible with any world that humans are capable of engineering. God might be seen as an explorer, prospector, perhaps a venture capitalist – or more theologically, a Moses who leaves it to others to inhabit the Promised Land. In a more contemporary guise, humans might be seen as divine nanobots, except that our programming allows for variation in performance. This means that we have the capacity for disobedience but God equally has the option to condemn or disown us. In this respect, the Clientelist God differs most strikingly from the Deistic God in being more like the engineer's savvy patron than the engineer himself. A relevant model for this relationship is principal–agent theory in political economy, whereby God is the 'principal' who employs humans as his 'agent' in a contract of mutually circumscribed trust. Specifically, God has agreed to trust our competence to get the job done but in exchange he cannot be held responsible should we fail, for that would be to abrogate both God's foreknowledge and our free will.

The **Ecological** model of God-playing is pagan in origin, based on the idea of an overall natural order that is identified as a state of equilibrium, a set point to which all changes eventually return, regardless of the efforts of the parties involved. Ancient Greek drama – both comedy and tragedy – is predicated on this idea. Accordingly, any attempt to play God eventuates in oneself being played. In Judaism, the threat of this prospect set the limits on retributive justice. It is also evident in the more conservative strains of Augustinian theology that stress humanity's naturally fallen state, which may tempt us to achieve god-like ends that are not properly within our reach that will only result in more misery. Consider William Paley, the early nineteenth-century natural theologian whose work contributed to Darwin's turn to atheism and is today treated as the godfather of the brand of scientific creationism called 'intelligent design theory'. Paley, while sympathetic to Deism's detached view of the deity, also supported fellow cleric Thomas Malthus's opposition to the 'Poor Laws' (an early income redistribution scheme) in order to provide a fair test of humanity's forbearance in the face of the misery and early death of the impoverished. This natural culling of the population resulted in a sustainable world for the worthy survivors, a doctrine that within 50 years Darwin would himself secularize as 'natural selection' and Herbert Spencer would popularize as 'survival of the fittest'. However, contrary to this entire Malthusian line of thought, improved public health in the late nineteenth century served to increase population numbers and longevity, leading 'racial hygienists', who practised a kind of ecological medicine, to warn against these mistaken 'counter selection' policies that favoured the 'unfit' (i.e. the poor, disabled, etc. who would otherwise 'naturally' face an early death) and would eventually strain resources and lead to global war (Fuller 2006: chap. 14). Nazism emerged as a populist downstream expression of this sentiment. However, concerns about human overpopulation, resource depletion and environmental despoliation persist to this day. Chemist James Lovelock (1979) coined the term 'Gaia' to capture the idea of the Earth as a superorganic deity, which serves to put a face on Darwinian natural selection, one that has been subsequently endorsed by the sociobiologist E.O. Wilson (2006). On this view, humans, while very much dependent on the Earth for their own survival, nevertheless are unique in their capacity to comprehend it whole – but only as long as they do not try to outsmart her. ('You can't fool Mother Nature!') At best, we are Gaia's physician, an idea promoted by the early twentieth-century biochemist Lawrence Henderson (1970) as a secular update of the Calvinist idea that we are God's stewards.

Finally, **Expressivism** accepts that life is a temporally extended cosmic accident, a success of self-organizing, self-reproducing processes that evolve into the autonomous systems we call organisms. In the course of this evolutionary history, humans emerge as the locus of cosmic self-consciousness: We have the capacity both to understand all that has gone on before and to resolve that complex legacy in a meaningful direction for the future. In this context, 'to play God' amounts to discovering our potential to become God, an entity who cannot be fully realized without our agreement to play the role. Our impulse to assume the divine position is evidenced in our ability to stand outside ourselves and adopt the standpoint of 'the view from nowhere'. As noted at the start of this chapter, this capacity is associated in the New Testament with the 'transfiguration' of Jesus into Christ and helps to explain the very existence of Deism and Clientelism. But it is also manifested in the seemingly endless frontier of scientific inquiry, which takes us far beyond what we need to know to sustain our normal physical existence. In this respect, Expressivism implies a kind of dynamic pantheism. After all, science destabilizes our existence on a regular basis, most noticeably in terms of transforming the selection conditions required for our own survival. Thus, the history of biotechnology from animal and plant husbandry to today's genomic manipulations extract from nature sustainable forms of life that would not have existed without our intervention. These then provide the ('smart') environments that condition our continued development. In short, taken as a piece of secular theology, Expressivism converts natural contingency into a 'miracle' by noting the extent to which an original event of self-consciousness created a path-dependency that has channelled the subsequent development of the cosmos. Kelly (2011) provides a purely technology-driven variant of this line of thought.

A couple of conceptual relationships among the four paradigms are worth noting. The Deist and Clientelist models are both anchored in an Old Testament sense of God and humans as engaged in some sort of communication whereby each knows what is expected of the other. Their difference lies in the implications of human autonomy for the divine project: For the Deist, the divine project is realized regardless of what humans do, whereas for the Clientelist human failure means that the project remains incomplete. For their part, the Ecologist and the Expressivist share the idea that humanity is literally part of God. But whereas the Ecologist treats God as a well-defined whole (the Earth) in terms of which we are a well-defined part (a species), the Expressivist sees God as the projected ideal future in terms of which humanity is defined as the ongoing vehicle of its realization, whatever form that takes. The above characterizations of the four theomimetic paradigms are meant to provide 'ideal types' for playing God that are present, in various combinations, in contemporary policies. In terms of the precautionary–proactionary ideological polarity introduced in Chapter 1, the Ecologist and the Expressivist, respectively, most closely approximate the difference at stake because both ends of the pole, however steeped in theology, are in fact enacted in the secular world, which is to say, we regard ourselves as either a part of an already fully functioning nature or a stage along a journey that results in our bring about a fully functioning nature.

3
Proactionary Biology: Recovering the Science of Eugenics

1 Transhumanism as Eugenics 2.0

If *The Proactionary Imperative* is so future-oriented, why are we so preoccupied with history? The answer is that history makes a progressive contribution to general culture by searching for decision-points in the past that set in motion a train of events that over time have resulted in something that we now take for granted. But the progressive historian's point is that things could have gone otherwise back then and can go otherwise here now. Thus, the proactionary argues that the current situation is sufficiently similar to that historic turning point to allow us now to pursue a path significantly different from the one chosen back then. It follows that a proactionary reading of history does not aim to legitimize the present but to reveal the original openness of the past – when what now appears inevitable was merely optional. Whenever history has 'critically' informed the contemporary scene, it has been on this epistemic basis, which normally travels under the rubric of 'counterfactual history' (Fuller 2008c). Proactionaries assume that there is enough freedom and flexibility in the human condition – if not the causal structure of time itself – that latter-day analogues to 'paths not taken' may be revisited anew (cf. Fuller 2010: chap. 9; Fuller 2011b).

The early twenty-first century resembles the early twentieth century vis-à-vis the openness of the boundary separating the social and natural sciences – more specifically, the disciplinary divide between sociology and biology. At stake is nothing less than what it means to be 'human', especially in a sense that requires a special body of knowledge somewhat set apart from the study of living things more generally (Fuller 2006). In other words, should the natural sciences be seen as allowing us to be more than our history has so far allowed us to be or, rather, as prescribing that humanity be subsumed as a marginal case of the normal workings of a nature that is fundamentally indifferent to our fate? In terms of competing conceptions of biological evolution, Lamarck and Darwin have stood for these positions, respectively. With the advance of scientific research, the choice in favour of a pro-human position has been made more difficult, since the empirical findings appear to support the Darwinian view that there is nothing special about *Homo sapiens* understood in strict biological terms that might permit it to control its own destiny any more successfully than the other organisms with which we cohabit the planet.

Love it or loathe it, eugenics stepped into the breach to address this problem, precisely in a manner that was designed to be favourable to humans. However, to be true to itself, eugenics requires mass surveillance and experimentation, with the understanding that many in retrospect may turn out to have been used or sacrificed for science, given what may be an irreducible uncertainty about how particular genetic combinations function in particular environments. It is easy to see why a strict Darwinist (including Darwin himself) would find such a proposition unpalatable. To such a person, eugenics looks like aiming to beat the house in a cosmic casino where all previous players have come away broke (cf. Fuller 2007a: chap. 3). However, the eugenicist presumes that s/he as a member of *Homo sapiens* knows something that prior species did not, which may be to do with our special ties to the cosmic casino's proprietor. It is not by accident that Jacques Monod's (1974) popularization of the metaphor of the 'cosmic casino' was adapted from the Christian eugenicist founder of modern population genetics, Ronald Fisher (1930: 37), about whom more below.

Unfortunately, our ability to consider these matters dispassionately has fallen foul of Nazi Germany, during which the project of eugenics that self-styled progressive thinkers had been promoting for the previous half-century, was 'nudged' towards a genocidal policy focused on the Jews and other 'undesirable' groups. ('Nudged' is used deliberately since many scientists embraced Nazism because of the greater freedom that it allowed to human subject experimentation, often without due diligence to the larger political context [Deichmann 1996].) Ethicists are often concerned with problems that stem from the wrong means being used to the right end. In the case of eugenics, we are faced with what may turn out to have been the right means but used to the wrong end. The practice of eugenics is not itself a moral problem. Rather, the problem lies in how this long-term, scientifically based social policy was used to achieve immediate political aims, which in turn has coloured how this policy prior to Nazism is understood. It is incumbent upon proactionaries to re-engage constructively with this history, since 'transhumanism' – at least in name – owes its very existence to eugenics, whose spirit it continues to promote under the slightly more politically correct rubric of 'human enhancement'.

'Transhumanism' is a coinage of Julian Huxley, grandson of Darwin's great defender, Thomas Henry Huxley. Julian is remembered today among fellow biologists as the British originator of evolutionary theory understood as (after the title of his 1942 book) the 'modern synthesis'. This is the combination of Darwin's original vision of natural history and Mendel's experimental approach to genetics that continues to anchor research in the more theoretical reaches of the biological sciences. However, social scientists perhaps better know Julian as the architect of the UNESCO 1950 statement on 'the race question', which appealed to science to establish an international juridical understanding of 'race' as a social construction (Brattain 2007). On the surface, these two achievements might seem to be in tension but in fact they reflected the overarching 'transhumanist' direction of Huxley's thought.

Huxley realized that we are the first species to understand how all of evolution works and hence capable – indeed obliged – to take control of the direction that it takes in the future. Of course, if Huxley is right (and we are inclined to agree with him), one might reasonably wonder whether *Homo sapiens* was fully 'human', either ethically or cognitively, prior to its having acquired such cosmic awareness. In any case, unlike today's Darwinists, Huxley had the nerve to suggest that we might assume moral responsibility for 'natural selection', given Darwin's discovery of its *modus operandi*. So, as we learn more about the molecular workings of heredity (and keep in mind that Huxley originally wrote before the DNA revolution), the question of whether members of *Homo sapiens* are allowed to exist in a state of free mobility or enforced segregation will become a political decision, since both extremes – as well as the intermediate states – will become increasingly realizable. What in the past could only be accomplished through largely self-organizing reproduction and migration patterns operating over many generations in the future might be achieved much more quickly by strategically targeted interventions involving both technology and legislation. The laboratory-like conditions of the agricultural station should not be underestimated as the platform for piloting schemes with such prospects in view – in effect, the time–space compression of evolution. On this basis, for example, Sewall Wright, the American founder of modern population genetics, demonstrated how evolution might be expedited by a policy of subdividing a group so as to allow a potentially beneficial mutation to incubate for several generations in isolation before being re-integrated with the rest of the group so as to shift its overall balance of traits.

When reflecting on the legacy of eugenics to transhumanism, it is worth recalling that as soon as Darwin's cousin, Francis Galton, launched the field in the 1860s, its capacity for bridging the emerging natural–social science divide was quickly recognized. On the European continent, it fed into the major role – rarely told to social scientists today – that *vangardiste* medical and education faculties played in promoting the nascent social sciences. Indeed, only a unique set of circumstances prevented the first chair in sociology in the UK from being awarded to a eugenicist. All but one of the candidates at the London School of Economics (LSE) in 1907 had espoused some version of eugenics. The outlier, L.T. Hobhouse, an Oxford-trained social philosopher of the then fashionable neo-Hegelian sort who wrote for the *Manchester Guardian*, was selected as Professor. Although the LSE's main benefactors, Sidney and Beatrice Webb, were clearly sympathetic to eugenics, a *de facto* chair in the field – held by Galton's follower and pioneer statistician Karl Pearson – had already been established at the University of London's flagship college. This bit of institutional politics probably did more to make a eugenicist a non-starter for the founding sociology chair than any antipathy to the eugenic orientation itself (Renwick 2012).

Keep in mind that even at the dawn of the twentieth century neither biology nor sociology was a clearly defined field anywhere. 'Biology', a coinage of the first modern evolutionist, Jean-Baptiste

Lamarck, challenged the classical way of thinking about 'nature' as consisting of animals, vegetables and minerals as three equal modes of natural being. Instead Lamarck drew a sharp ontological distinction between living and non-living matter, effectively establishing the disciplinary boundary between biology and geology within the field that had been recognized from Aristotle to Linnaeus as 'natural history'. Fifty years later, Darwin was already taking this distinction for granted, as he tentatively posited the 'primordial soup' as part of an atheistic account of the transition from non-life to life. But Galton did not see his science of eugenics as an application of biology to human affairs. Rather, he saw it as an extension of political economy to its final frontier, which is to say, the conversion of humanity to capital, or as we say nowadays, 'human capital' (Renwick 2012: chap. 2).

Nineteenth-century political economy was a quasi-normative discipline that treated everything as capital that could be inherited, accumulated, enhanced and transmitted. In terms of its ontological reach, one might think of political economy as the social science that corresponds to a chemist's way of seeing the human condition, just as psychology might be seen as the social science that corresponds to how the physicist sees us. Political economy and chemistry are both largely concerned with exchange or translation relations between raw material and meaningful entities. Against this backdrop, the incorporation of the human under the category of capital would finally give political economy theoretical closure. The barrier in the way has been the *labour theory of value*, a medieval idea based on humanity's divine inheritance that was enforced by natural law. To be sure, contrary to Marxists' wishes, labour has been eclipsed by exchange as the source of economic value over the past 150 years. Nevertheless, a good indication of capitalism's lingering attachment to the biblical mystique of the human form is that normative regimes in capitalist societies have historically fretted the most about enslavement, either through employment or marriage – though the grounds for concern have been always ambiguous between systemic concerns about the efficient flow of capital and the intrinsic rights of individual humans (Collins 1999: chap. 6).

It is against this context that eugenics exploited the bio-capital implications of the legal idea of 'inheritance', which cannot only be taxed but also, so to speak, bred and farmed. Political economy had come into its own as what Marxists deride as the 'science of capitalism' only once it junked the eighteenth-century French physiocratic idea that land – as proxy for nature – was the source of all value, and focused instead (chemistry-style) on a conception of value as the conversion rate between forms of capital. At that point, political economy became committed to indefinite growth through ever more efficient substitutions of natural by artificial means of production, resulting in ever more productive forms of capital. In this context eugenics may be understood as extending the idea of increased agricultural productivity to what Darwin's French translator, Clémence Royer, called *puériculture*, which takes the idea of 'raising children' to a new degree of literalness (Hecht 2003). When the political economy backdrop to Galton's thinking is kept in view, then the route from late nineteenth-century eugenics to early twenty-first-century transhumanism is made clearer.

Perhaps the founding moment of this development was *The Principles of Political Economy* (1817) by the English stockbroker, David Ricardo, a Sephardic Jew who converted to Unitarianism, the dissenting Christian sect dedicated to human self-empowerment that in his day was associated with the radical chemist Joseph Priestley (Fuller 2011a: 196–201). Whereas Ricardo's older contemporary, Thomas Malthus (himself schooled in Priestley's curriculum at Warrington Academy), still believed that nature places an outer limit to productive growth, Ricardo abandoned that assumption, recognizing that even human labour would gradually lose its value through the introduction of more efficient mechanical substitutes.

In this respect, it is rather misleading to follow the normal practice of including Ricardo with Adam Smith – and then John Locke and Thomas Aquinas – as supporting a 'labour theory of value'. The spirit of the labour theory of value, which survived in the Marxian corpus, is connected to a strongly normative sense of 'natural law', according to which human labour possesses absolute value, which is the source of the idea of 'just wage'. The quantity of labour was not supposed to be abstracted from one's being a labourer, which is exactly what Ricardo did. For him, the value of labour lies in the amount needed to make a commodity, *regardless of who or what delivers it*. Ricardo's 'labour' is not a constant but a variable – one normatively spun in the direction of 'least effort'. It truly reduces 'human labour' to a physical capacity for work. The 'humanity' of this process, one might conclude, lies in the very act of exchange, as two (nominally) free parties agree to a conversion of capital. In terms introduced in the previous chapter, Ricardo must be counted as an 'ephemeralist' (and perhaps even the patron saint of transaction economics), as he regarded markets as sacred spaces where people perform as Maxwell's Demon, such that the work involved in agreeing an exchange results in increased efficiency in the performance of both traders.

Once Ricardo got his way in political economy, the door was opened to make all sorts of previously unseemly comparisons: e.g. one well-paid worker who dutifully works on schedule *versus* many poorly paid workers whose erratic performance collectively produces more. This prospect is worth keeping in mind when considering the motivation for introducing 'minimum wage' laws in the early twentieth century. The Marxist-style trade union argument for a minimum wage law pertained specifically to the welfare requirements of the humans who are expected to return to work and do a good job on a daily basis. However, the Ricardian argument for the same policy – common to the British Fabians and the American Progressives – focused instead on the poor quality of workmanship that would result for the manufacturer (and potential consumers) who allowed workers to 'rush to the bottom' in terms of the wages at which they would be employed (Leonard 2005). But there is an unintended consequence of workers having to be hired at a wage higher than the market rate: namely, manufacturers become more discriminating of their prospective employees, while the workers, still operating in the midst of many market rivals, are motivated to demonstrate more value for the extra wage that their potential employers would be having to pay them. In effect, then, minimum wage laws raise the collective intelligence of the labour market by forcing everyone to play smarter, not simply cheaper. More than their contemporary Adam Smith, the French Enlightenment philosophers Turgot and Condorcet had recognized this instructional role that markets might play, especially in the hands of a benevolent state (Rothschild 2001: chap. 7).

Ricardo himself appeared to have believed, as neo-liberals do today, that this situation also provides an incentive for workers to acquire smarter skill-sets, if not commit themselves to 'lifelong learning', to keep up with the market. In response, Marx observed that this Ricardian vision of capitalism seemed 'inexorable' only if the laws of political economy followed the path of least resistance to the capitalist employer. Ricardo's science of capitalism was in reality a science *for* capitalism. As a good Unitarian, however, Ricardo could try to regain the moral high ground by saying that Marx underestimates humanity's capacity for individual self-transformation. To be sure, Marx was rhetorically effective in mobilizing workers to organize themselves and speak with one voice, but it was at a cost. He reverted to the labour theory of value associated with the natural law tradition, even though his own historical materialist metaphysical framework did not support it. Marx clearly did not want to turn back the clock to pre-capitalist days, since the efficiency savings encouraged by the capitalist mode of production was a necessary condition for a Communist paradise. Nevertheless, unlike Ricardo, Marx shared the natural law theorists' commitment to the integrity of the paradigmatically 'normal' human body, the legacy of which remains in the pejorative tinge attached to 'exploitation'. However, in practice, successful self-styled 'socialist' governments – be they in Scandinavia, Germany or Russia – operated in a more Ricardian spirit than Marx would have wished, one favourable to eugenics.

Galton's relevance to this debate is complex. While Galton questioned Ricardo's faith that individuals have the wherewithal to acquire new traits in response to changing market conditions, he refused to concede the finality of Darwin's Malthusian tendency to view these market shifts as expressions of natural selection that effectively decide who is fit to live. At the same time, Marx's counter-strategy struck Galton as relying on an outmoded, even fetishized view of human labour (of the sort promoted by the medieval guilds) that failed to distinguish socially desirable traits from those who happen to bear them at a given time – a distinction Ricardo had clearly recognized. Galton's own strategy was to take the long view and try to persuade people that society's desirable traits are not normally well distributed across living individuals. Nevertheless, this sub-optimal situation may be remedied by proactive policies designed to encourage and discourage births of certain sorts.

Precedent for this move could be found in the mentor of positivism's and sociology's founder, Auguste Comte: Count Henri de Saint-Simon. Saint-Simon, unfairly dubbed by Marx and Engels as a 'utopian socialist', subsumed the human body under the category of 'property', the rational administration of which requires collective ownership and expert management (Fuller 2011a: 142–6). In that case, personal autonomy should be seen as a politically licensed franchise whereby individuals understand their bodies as akin to plots of land in what might be called the 'genetic commons', subject to all the rights and duties implied by the analogy. We address this matter under the rubric of 'hedgenetics' in Chapter 4. An open question: does this 'genetic commons' correspond to a racialized nation-state, a global human species, or perhaps some open-ended pool of all genetic material in relation to which 'humans' function only as recognized 'legal persons', that is, as bearers of rights and duties regardless of heredity? The last would be truest to the current state of biological knowledge (cf. Oldham et al. 2013). However specified, the ultimate goal in this bio-capital utopia is maximum productivity – making the most out of one's inheritance. To be sure, 'irrational' (aka traditional) socio-economic barriers are likely to prevent some individuals – especially of poor backgrounds – from achieving this goal. And while wealth redistribution and egalitarian legislation might well address much of this problem in the short term, a more comprehensive long-term solution requires improving the capital stock of humanity itself. So goes the logic that leads to eugenics.

This last point is worth stressing for two reasons. One is contemporary: when faced with the shortfalls from the redistributivist and egalitarian policies that Western social democracies have pursued since the 1960s if not earlier, it is nowadays common for Left-leaning, biologically minded thinkers to declare – as Darwin himself might – that there are definite limits to how much people can be changed. Indeed, Peter Singer (1999) went so far as to advise his fellow Leftists to ditch Marx for Darwin. Whatever else one might wish to say about eugenics, it did not give up so easily – or more precisely, it had a more consistent faith in the import of new knowledge (aka 'basic research') for future policymaking. The other reason to elucidate the logic behind eugenics is to dispel a pervasive historical stereotype. Because eugenics continues to be closely associated with the totalitarian regime of Nazi Germany, Galton's science is often seen as aiming for policy outcomes much more quickly than could (or would) be achieved by normal democratic processes. However, in the British soil where eugenics first took root, its most outspoken advocates – Sidney and Beatrice Webb – identified themselves as 'Fabian socialists', in the spirit of Fabius, the Roman general who refused to act impulsively against Hannibal in the Punic Wars but nevertheless won in the end. In other words, eugenics was supposed to provide a blueprint for basic research in the social sciences with a rather long time horizon, comparable to the experimental turn that had enabled the natural sciences to break with the natural history tradition over the previous two centuries.

Recall that the people normally taken to be the founding fathers of the social sciences (excluding psychology) believed that we either already knew enough about the human condition to now focus on its political implications or, if our basic knowledge was still lacking, we would proceed more systematically but in the largely comparative-historical mode of traditional humanistic scholarship. The former category included Comte, Mill and Spencer, while the latter included Durkheim and Weber, with Marx believing in a bit of both. Galton's eugenics was arguably the first discipline to offer a clear statement of a basic research programme for the social sciences that had something like the character and dimensions in terms of which funding agencies think about such matters today – that is, a strong theoretical framework operationalized in terms of clear methodological strictures that enabled the collection and analysis of a wide range of original data. Put this way, it should come as no surprise that Otto Neurath, the sociological founder of logical positivism, was Galton's German translator. Indeed, eugenics would not have been such an easy target for censure, had it not set its own scientific standards so high – something for which the field has yet to be given due credit. As we shall see in the next section, the epistemic significance of eugenics was not lost on the father of the British welfare state, the economist William Beveridge. In 1930, as director of the LSE, Beveridge hired the experimental biologist Lancelot Hogben to establish a department of 'social biology' that would provide a 'natural basis for social science'.

2 Recovering biology's lost potential as a science of social progress

The British journalist Jonathan Freedland (2012) recently described eugenics as 'the skeleton that rattles loudest in the Left's closet'. He wrote as someone who finds this noisy dwelling his natural ideological home – as do the authors of this book. Freedland's article includes a photograph of the founder of the British welfare state, William Beveridge, with the caption: '[Beveridge] argued that those with "general defects" should be denied not only the vote, but "civil freedom and fatherhood"'. Admittedly Beveridge wrote these words in 1909, a decade before assuming the directorship of the London School of Economics and more than three decades before issuing his famous report that established the UK welfare state. Nevertheless, though barely 30 at the time, Beveridge attracted the eye of both the Fabian Society and the equally precocious Winston Churchill, who appointed Beveridge to the UK Board of Trade where he drafted a prototype for a national insurance scheme aimed at the unemployed. Moreover, as Chris Renwick (2013) reminds us, the idea that a 'social biology' might provide the foundation for both the social sciences and a scientifically informed welfare policy remained integral to Beveridge's world-view throughout his life (cf. Sewell 2009: chap. 3). But is this fact a mere historical curiosity with no larger normative significance for our own times? Or does it speak to a deeper affinity between eugenicist and welfarist thinking that merits revisiting by those interested in reinvigorating the welfare state in the twenty-first century? We defend the latter proposition.

No historically literate person can deny the roots of welfarism in broadly eugenic considerations – not only in Britain but also Scandinavia, Germany, the United States and elsewhere. This fact has been seized upon by both classically liberal and culturally conservative commentators as evidence for the depravity that lies at the heart of any politics touched by socialism. Indeed, *National Review* columnist Jonah Goldberg (2007) topped *The New York Times* non-fiction best-seller list in early 2008 with a book entitled 'Liberal Fascism'. It aimed a torpedo at the presidential hopes of then front-runner Hillary Clinton, who had drafted an ill-fated national health insurance scheme when her husband occupied the Oval Office. With all the finesse of a high school debater, Goldberg laid down a comprehensive bill of charges that, despite much nitpicking by critics, clearly showed that Beveridge's eugenic welfarism was hardly idiosyncratic, but shared many of the same concerns and policy orientations as 'national socialist' movements of the same period. To be sure, Goldberg wanted his account to provoke the Left and shock the undecided in the upcoming 2008 US elections. Nevertheless, more enduring questions remain, most importantly: can a welfare state today be countenanced without dealing explicitly with the biological side of social life to which eugenics drew such vivid attention?

The distinctive cast of Beveridge's welfare state reveals his particular take on eugenics, one common to the Fabian Society, the early twentieth-century vanguard movement that took its name from the great patient Roman general who defeated Hannibal by letting the latter's forces wear themselves out in their approach to Rome. This precedent was meant to do double work for the movement, the first involving vigilance and the second perseverance.

On the one hand, Fabius reminded them that what Graham Wallas had christened (and Lyndon Johnson would later adopt as his own) 'The Great Society' would not be accomplished in a single political cycle. It would require a regularly monitored, long-term strategy that was responsive to events but kept the ultimate goal in view. Thus, the Fabians are rightly contrasted with, say, the Bolsheviks in terms of advocating a 'non-revolutionary' brand of socialism. However, it would be a mistake to see them as especially friendly to parliamentary democracy. Rather, the 'non-revolutionary' means they had in mind involved the persuasive power of science – specifically, of eugenics. This would require the commissioning of mass longitudinal studies that brought an unprecedented level of surveillance into households and workplaces, carefully dividing people into race, class and gender. Thus was laid the groundwork for what is today called 'quantitative sociology'. Results from such studies would be regularly fed into the legislative process to influence ongoing debates; hence the Fabian Society's fair reputation as the original think-tank.

Meanwhile Fabians such as G.B. Shaw, H.G. Wells and Julian Huxley provided a steady stream of pro-science propaganda designed to enable people to integrate the emerging eugenicist world-view into their lives. Nearly a century later, this genre of literature, which the Fabians coined as 'New Age', has been replaced by social scientifically inspired focus groups and wiki-media designed to acclimate people to various 'nano-bio-info-cogno enhanced' futures. It is practised under the rubric of 'anticipatory governance' in the field of science and technology studies (Barben et al. 2008). But unlike the Fabian initiative, it follows rather than leads the science policy agenda (Fuller 2011a: 147). Undoubtedly this reflects today's researchers surviving on short-term, university-based contracts rather than on the largesse of private foundations. In any case, the Fabians knew (in the sense of 'owned') what they were doing, a lesson from which today's 'neo-liberals' benefit after some Fabians crossed the aisle 75 years ago, inspired by the US journalist Walter Lippman, to embark on a new long-term strategy for global ideological domination, one marked by a greater distrust in democracy and a greater faith in enterprise than even the original Fabians had: the Mont Pèlerin Society (Plehwe 2009).

On the other hand, the image of the patient Fabius was also meant to convey an air of objective, perhaps even cold detachment in the face of an oncoming enemy – not reticence, or even cowardice, as the Fabians were sometimes portrayed by socialists who were more given to bold, violent gestures. This sense of detachment was reflected in Beveridge's particular construction of the welfare state, which stressed what we now call 'equality of opportunity' but not 'equality of outcomes', a feature that distinguished the UK from Scandinavia and more 'truly' socialist regimes (Benassi 2010).

Here it is worth recalling that socialism evolved from two features of industrial capitalism that David Ricardo's stripped-down analytic treatment had rendered problematic for defenders of the system (Gordon 1991: chap. 9). The Marxists picked on the labour theory of value, which Ricardo accepted but then treated as a variable the exact value of which employers would always try to minimize by finding cheaper, including machine-based substitutes. Ricardian rationality thus morphed into Marxist injustice. But the Fabians were less interested in setting a 'fair' or 'natural' price for labour than in ensuring that 'capital' in the broadest sense – one that anticipated the neo-liberal rendering of 'labour' as 'human capital' – was utilized with maximum efficiency (Becker 1964). It is here that the rhetoric of 'a life wasted because of lack of opportunity' belongs. In this context, the Fabians generalized Ricardo's own antipathy to the very idea of *rent* – that is, the capacity to derive income from unproductive ownership. Here they were influenced by the US economist Henry George's proposal for a single tax on unused land, which was aimed at the idle rich, who were just as much a threat to national prosperity as the idle poor (Goldberg 2007: chap. 7). This side of 'idleness', explicitly targeted as one of the 'five giant evils' that Beveridge's conception of the welfare state was designed to tackle, remains underexplored. (The relevant US comparator is the Progressive Yale economist Irving Fisher, who explicitly referred to humans as 'capital' – without irony or condemnation – as early as 1897 and later became the first president of the American Eugenics Society.)

An intuition that links Ricardo and the Fabians, via Francis Galton, is that any sort of 'inheritance' – be it based on a legal or a biological definition of 'descent' – is so *prima facie* suspect as to constitute a secular version of Original Sin. (The ideal case would be a being whose productivity comes completely free of debt to the past, someone capable of creating everything out of nothing, the proverbial 'self-made man', as per Augustine's account of divine creativity as *creatio ex nihilo*. This is the kernel of goodness that lies buried beneath the ambient nastiness of Ayn Rand and her libertarian followers.) Here it is worth recalling Galton's policy motivation for eugenics lay in reforming the hereditary House of Lords, in which generations of descendants received a free ride to power and privilege on the back of one distinguished ancestor. In terms of the underlying normative intuition here, consider that Auguste Comte's neologism, 'altruism', was designed as an ideological antidote to just this state-of-affairs. Comte saw each human as born into debt – paid out in terms of germ plasm and (when fortunate) property – that can only be redeemed by conducting one's life in a 'pay-it-forward' mode (Graeber 2011: chap. 3). In short, what in the language of medieval essentialism would have been called 'realizing one's potential' was now historicized as 'repaying one's debt'. This is the context in which Rudyard Kipling's phrase 'white man's burden' makes sense, which in turn explains the Fabian partiality to imperialism. Indeed, Beveridge himself was a child of the Raj.

However, exactly who is likely to benefit from one's repaid debt remained a tricky question, since the nature of the capital that constitutes the debt has both an 'internal' (genetic) and an 'external' (economic) dimension. Thus, there would be need for guidance on consumption along both dimensions. To put it in the blunt terms that John Maynard Keynes – a Fabian-friendly opinion leader in both finance and eugenics – would have recognized: the less genetically indebted (i.e. 'regressives') should live an entirely self-consuming existence without remainder (i.e. no savings and no offspring), while the more genetically indebted (i.e. 'progressives') should be encouraged to invest their money and germ plasm in ways that are likely to bring the greatest long-term return. Dividing the national portfolio of human capital in this fashion provided the best strategy for increasing overall prosperity, as the poor enjoyed themselves in the present and the rich built for the future. (Shades of Aldous Huxley's *Brave New World.*) Over the course of generations, then, a society of autonomous individuals would be bred: *natural born liberals*, if you will. In this respect, socialism's collectivization of the 'means of production' – understood in this broad 'human capital' sense – appealed to the Fabians not as an end in itself but as the most efficient means to cultivate a race of liberals (Wheatcroft 2012).

The 'only' problem was how to identify the 'rich' and 'poor' in a way that enabled eugenics to breed and select liberals in the name of human capital development. After all, Galton's original insight was that the distribution of wealth in Victorian Britain was more closely correlated with reproduction than production – that is, more backward- than forward-looking. Against this backdrop, the call for 'equality of opportunity' was designed to turn society into a mass living laboratory in which everyone had a fighting chance to display their hidden wealth, which would then be subject to a genetic audit, aka periodic nationwide examinations. In this respect, universal health, education and child-care benefits aimed to level the playing field in preparation for exams whose results would contribute to the streaming of students into the most productive life course, resulting in a 'meritocracy'. Unfortunately, political ambitions exceeded the reach of science. Knowledge of genetics, though advancing rapidly after 1900, had already shown the difficulty of determining the full range of one's genetic potential based on studying family histories, let alone the likelihood that specific individuals will display specific traits. This was fully appreciated by the Fabians, which explains why a sophisticated methodologist, Lancelot Hogben, was an ideal choice for the chair in social biology.

From his memorable inaugural lecture in 1930, Hogben made it clear that he did not intend to impose the sort of biological imperialism that came to be associated with, say, E.O. Wilson's 'sociobiology' in the 1970s. On the contrary, Hogben agreed with Beveridge that the uncritical extension of animal-based studies to human populations is profoundly unscientific, making for capricious policy. As Hogben wittily put it, social biology needs to be less about 'the sterilization of the unfit' than 'the sterilization of the instruments of research before operating on the body politic'. The shorter Hogben: Sterilize your own mind before you sterilize others' bodies. Thus, Hogben was inspired to adopt a profound re-reading of his subject's history, which led him to recover human genetics (then called 'eugenics') as 'political arithmetic', one of the Royal Society's founding concerns, when political economy and biology were unified in a statistically based science of population management placed in the service of nation building (Hogben 1938).

Importantly, Hogben re-read this concern using Marx and Mendel, not Malthus and Darwin, as the main conceptual turning points. The result was a stress on the prospect for movement both within and among people, rather than the discovery of where people belong if they are to survive on their inheritance. Here it is worth recalling that Hogben's science – population genetics – moved freely between speaking of 'population' of genes and of organisms, understood as the bearers of genes, which often gave the impression that the migration of semen and souls was interchangeable. Indeed, Hogben's own lifestyle as a free-spirited cosmopolitan personified this theoretical perspective, one that treated with suspicion any notion of a genetic 'homeland', be it Nazi or Zionist in inspiration (Werskey 1978).

However, a question that has continued to haunt welfare states is the extent to which Hogben's liberal, even experimental attitude towards the national gene pool is fiscally sustainable. Hogben's own justification was the increasing levels of dislocation and death that he thought was likely to result from – to put it in Cold War parlance – the 'military-industrial complex'. In short, the perpetual prospect of de-population provided an incentive for genetic innovation (Fuller 2011a: 39–44). Thus, Hogben contributed to the steady transformation of Dr Pangloss to Dr Strangelove (or, in world-historic terms, Leibniz to Herman Kahn) in the secularization of theodicy, which was raised in the previous chapter and will be discussed again in the next section (see also Fuller 2010: chaps. 1, 7; Fuller 2011a: chap. 5).

Unfortunately Hogben's tenure was dogged by discontent from the LSE economists, notably ones who would be foundational for neoliberalism, Lionel Robbins and Friedrich von Hayek (Plehwe 2009). Beveridge never quite got a handle on the problem, which went beyond ideological opposition to any form of *dirigisme* to a deeper scepticism about the epistemic value of original empirical research aside from politically motivated surveillance. As for Hogben, he carried on regardless, until the Rockefeller Foundation, his principal funder, fatefully shifted its priorities from population genetics to molecular biology, which then in less than two decades resulted in the DNA revolution. That shift in resources, along with the scars left by the Nazi genocide, shelved indefinitely Hogben's emerging revisionist view of genetics.

In today's terms, and in the spirit of Beveridge's social biology, Hogben interpreted genetics as a hybrid natural-social science. Take the two central concepts of genetics: 'genotype' corresponds to 'natural capital' and 'phenotype' to 'social capital'. In that case, the 'environment' is understood in the broadest sense (i.e. from diet to community) as the platform for translating the former into the latter. Hogben proposed to tackle the various loci of intervention available for effecting desired translations. His agenda called for a complex of mass observation surveys, longitudinal site studies and laboratory experiments, much of which was launched in his tenure and remain as paradigms of social policy research – albeit without Hogben's overarching vision. Nevertheless, in his brief and unhappy tenure at the LSE, Hogben managed to launch a sophisticated survey of 4000 twins of school age in the London area to examine in some detail the relationship between heredity, environment and intelligence – with an eye to checking the validity of psychological testing. In the end, while Hogben failed to implement his own and Beveridge's prospectus for social biology, he nevertheless managed to train David Glass, who went on to become the doyen of British quantitative sociologists in the postwar era.

While population geneticists in the second half of the twentieth century increasingly recognized the multiple 'levels of selection' implied in Hogben's research strategy, the experience of the Second World War inclined them to treat it more as a deep conceptual problem than a merely technical engineering one. Where Hogben and his contemporaries might have wanted to conduct tests in order to learn from error, politically correct biologists backed by the scholasticism of analytic philosophy have subsequently discouraged such moves as violating the genome's 'irreducibly contextual' nature (e.g. Sober and Lewontin 1982). Under the circumstances, the only thing that kept genetics from marching back into the Dark Ages was the Rockefeller-funded revolution in molecular biology, which put 'biotechnology' on an entirely different footing – one much more open to trial-and-error approaches – that bypassed the self-loathing of the population geneticists.

The general Rockefeller strategy was to lure people who could bring new skills to work on standing problems – in this case, physicists and chemists to unlock the hidden sources of human potential, which from the outset was understood in ways that would enable their re-engineering (Rasmussen 1997). The founding inspirational lecture for this movement was the physicist Erwin Schrödinger's (1955) 'What Is Life?' It was followed by the chemist Walter Gilbert's (1991) manifesto for bioinformatic adventurism, which reads like a job specification for Craig Venter. Indeed, according to one of Gilbert's former students, now a leader in synthetic biology, the publicly funded US Human Genome Project came into being in response to the *Next*-like threat posed by Gilbert when he announced plans to sell genetic information that had been sequenced by his start-up firm BioGen in order to recapitalize the company in the wake of 1987's 'Black Friday' crash on Wall Street (Church and Regis 2012: chap. 7).

Before moving on, let us make explicit the point of engaging in the sort of 'counterfactual history' illustrated in our consideration of Renwick's 'paths not chosen' approach to understanding the trajectory of twentieth-century British social science. Such a history examines not only what did happen but also what could have but failed to happen – especially with an eye to resurrecting those lost possibilities as future opportunities. We can then appreciate Beveridge's failure to establish social biology as the foundational social science discipline at the LSE as more than a mere historical curiosity. An opportunity that had been lost at the end of the 1930s might be recovered today, given the current configuration of events. After all, historians, even if they are loath to admit it, do not presume that every event is equally necessary for every other event. For example, on Renwick's (2013) telling, Hogben was undermined by local factors, while more global factors militated against his agenda being resumed by others elsewhere after the Second World War. But all of these factors – with their short- and long-term effects – are no longer with us, while the issues that animated the original Fabian agenda clearly remain, at least insofar as we are interested in pursuing 'progress' as an explicit policy goal.

The more immediate question is how the historian can contribute to taking the discussion forward. My own construal of counterfactual historiography involves calibrating intuitions between the past and the present (Fuller 2008c). To cut a much longer story short, the scepticism that Hayek and other LSE economists originally expressed about social biology would be re-invented today by social constructivists who see science as simply one among many sources of knowledge to which individuals in the market might turn for guidance, but without enjoying any special privilege. However, given that biotechnology is nowadays a market phenomenon, any revival of Hogben's vision will require that academics re-establish their epistemic prerogative over the situation, perhaps as some state-authorized testing ground or market regulator (cf. Fuller 2000: chap. 6). However, this is unlikely to happen as long as academics endorse the research strictures laid down by 'institutional review boards', which only gives them less scope for action than that of other market players.

3 Against the 'wisdom of nature': Why transhumanists need to get over Darwin

At the level of rhetoric, and often substance, contemporary transhumanism presents itself as advancing, if not accelerating, 'evolution'. (The term is put in scare quotes to encourage the reader to query the exact process that we are supposed to be talking about.) Thus, a leading work of transhumanist bioethics is called *Enhancing Evolution* (Harris 2007). Moreover, transhumanists do not mean 'evolution' as a mere synonym for 'development' but rather as a process that can legitimately draw on Neo-Darwinism for support, not least as an account of how natural history might bear on our future prospects. Thus, the most well-articulated transhumanist manifesto bears the title, 'The wisdom of nature: An evolutionary heuristic for human enhancement' (Bostrom and Sandberg 2009). We shall argue that contemporary Neo-Darwinism, under its official philosophical interpretation, goes against the spirit of transhumanism, largely because the theory does not ascribe to *Homo sapiens* any species-transcendent, let alone global, capacity to control the process of natural selection. To be sure, Darwin modelled natural selection on the long-standing benefits of artificial selection – that is, animal and plant husbandry – but he was well aware of the creative limits of these processes.

The person of Charles Darwin remains a talisman for people who have lost their original faith in the face of scientific evidence yet without being converted to another faith. Wishful thinkers – not least in the transhumanist community – like to interpret Darwin's doxastic trajectory as pointing to a sort of autonomy to which even non-Darwinists aspire. But then Kant kicks in. To be freed from constraints previously placed on your will implies simply that you take responsibility for the constraints to which you now submit yourself. It does not follow that you will see the world as affording more opportunities – let alone the transhumanist's endless horizons. Indeed, the world may seem to provide fewer opportunities, as in Darwin's own case. The ethos he had inherited from his nonconformist Christian parents, which in his younger years led him to rail against slavery and racial inequalities, was replaced in later years by a *de facto* atheism that served to depress his sense of self and lower his expectations for humanity more generally (Desmond and Moore 2009). Darwin's pessimistic reading of the evidence was based on his belief that the burden of the past – the legacy we share with the other animals – would ultimately prove too great to overcome.

Transhumanists should regard Darwin as someone who fell at the final hurdle of the Enlightenment project, at least when it came to the full redemption of humanity's divine entitlement. This was the hopeful message of Jesus that the Enlightenment wished to deliver through the advancement of science but which 'Christendom' as the institutional expression of Christianity had so far betrayed, most singularly in the trial of Galileo. *Pace* Darwin, we cannot simply be capable of adaptively responding to changes in the natural environment. In addition, we must have the courage to adopt the role of Natural Selector. Darwin consistently shied away from this species-based arrogation because natural selection for him was ultimately an Epicurean deity whose omnipotence is matched only by its indifference to humanly designed ends. By the standards of his contemporaries, Darwin must be counted as a pessimist with regard to humanity's capacity to beat the odds posed by natural selection. Whatever else one might wish to say about Darwin, he advised *against* the larger ambitions that informed eugenics, vivisection and even contraception. Francis Galton, Thomas Henry Huxley and John Stuart Mill may be called, respectively, as witnesses to testify to this fact. Were Darwin to be teleported to our times, he would have some clear views about the neologism, 'anthropocene' – that is, the most recent geological era, whose origin ranges from 8000 to 2000 years ago, during which humans have been the main source of global environmental change. He would side with the precautionaries who see it as a harbinger of future catastrophe, not the foundation on which humanity shall build a 'Heaven on Earth'.

In stark contrast, transhumanists' exceptionally high regard for humanity's scientific potential is based on a judgement that our long-term but often unintentional species culls and large-scale environmental restructurings have left us in a very strong position to steer natural selection explicitly to our species advantage. An example of this optimism is Bostrom and Sandberg's somewhat convoluted yet scientifically informed discussion of the sickle-cell gene, the 'heterozygous' version of which makes humans resistant to malaria but the 'homozygous' version susceptible to sickle-cell anaemia:

> The 'ideal optimum' – everybody being heterozygous for the gene – is unattainable by natural selection because of Mendelian inheritance, which gives each child born to heterozygote parents a 25 per cent chance of being born homozygous for the sickle-cell allele. Heterozygote advantage suggests an obvious enhancement opportunity. If possible, the variant allele could be removed and its gene product administered as medication. Alternatively, genetic screening could be used to guarantee heterozygosity, enabling us to reach the ideal optimum that eluded natural selection. (Bostrom and Sandberg 2009: 401)

The tenor of this passage suggests that human ingenuity is well on its way to beating natural selection at its own game. This strategic mode of thinking runs deep in the essay, which describes 'evolution' as a 'surpassingly great engineer', the sort of intelligent designer who would not have been out of place in eighteenth-century natural theology, including the 'evolutionary optimality challenge' that such a deity poses to our attempts to outsmart 'the wisdom of nature' (378). This challenge consists in rearranging the various trade-offs that the divine engineer had to make in order for ours to be the best possible world overall, even if that required that the world be less than perfect in each of its proper parts.

Although Bostrom and Sandberg appear to be innocent of theology, they are approaching God's *modus operandi* very much as the great post-Cartesian philosophers Malebranche and Leibniz did under the rubric of 'theodicy', the science of divine justice, which in eighteenth-century Europe morphed into political economy – in Britain, via the radical Joseph Priestley, the conservative William Paley and their intellectual offspring, Thomas Malthus (Fuller 2010: chap. 7). Like these natural theologians, Bostrom and Sandberg conceptualize evolution's optimality challenge in terms of several biological demands vying for realization within the constraint of a common resource base – put concretely, the design of an organism for living in a particular environment. This understanding of 'nature's economy' had already informed Linnaeus's naming of the species by their functions, prior to its inclusion in the original definition of 'biology' laid down by the first modern evolutionist, Jean-Baptiste Lamarck (Bowler 2005). Darwin showed that these design features of organisms can be explained by the operation of natural selection on genetic variants over many generations without any need for an intelligent designer. He effectively reduced talk of 'optimality' in the order of nature, if done knowingly, to poetic licence (say, in the aid of comprehension, as in Richard Dawkins' [1986] 'blind watchmaker') or, if done sincerely, to a narcissistic delusion (as if species might be designed specifically so that we might make sense of them).

Nevertheless, Bostrom and Sandberg instruct the reader to treat the idea of evolution- as-divine-engineer as a metaphor worth stretching to see how much light it shines on the prospects for human enhancement. But to remain true to Darwin, this must be an exceptionally limited metaphor, given the radical difference in the underlying causal mechanisms associated with the metaphor's two sides. After all, from a strictly Darwinian standpoint, the appearance of design in nature is simply an artefact of whatever stability exists in the environment that enables an ecology of organisms to flourish for several generations. However, over time, the organisms may become too well adapted to the environment, such that even a slight shift in living conditions may trigger a dislocation of the ecology, resulting in mass extinctions. Moreover, this scenario is just as applicable to humans as to any other species. Indeed, by ruthless Darwinian logic, we might expect that *Homo sapiens* will meet a tragic fate, whereby a set of traits that privileged our existence on Earth (at least to our own satisfaction) over a certain period ends up being the source of our species downfall.

Darwin himself seemed to think that our hypertrophied cerebral cortex, with its capacity for fixed ideas and inflated self-regard, was a good candidate for the species killer. In short, big brains make for big egos that too easily amplify the significance of success and repress the memory of failure. Thus, given the cosmic indifference of natural selection, our ever expanding and fetishized brains may turn out to be a cancer eating away at the human superorganism. It is thus easy to see why Darwin had little time for his cousin Galton's preoccupation with identifying and breeding geniuses, since such people are arguably the cancer agents. From that standpoint, the very idea that we might acquire a second-order, systemic understanding of the entire evolutionary process – perhaps transhumanism's most fundamental epistemic assumption – is a collective psychic disorder symptomatic of the cancer's onset.

With touching arrogance, Bostrom and Sandberg turn these gloomy Darwinian prospects for our long-term survival on their head, which then gives them the confidence to dispose of 'the wisdom of nature'. Following the received view of today's evolutionary psychologists, they begin by noting that our bodies and minds are still designed for life 40,000 or more years ago (i.e. before the advent of writing and other space–time-binding social technologies) even though in that time *Homo sapiens* has managed to parlay its genetic capital to achieve things that have radically altered our existential horizons. In particular, the natural environment has been made 'smarter' in various ways – amounting to the world becoming a more efficient place in which to operate – that has allowed us to realize many of our wildest dreams. Indeed, such is technology's grand world-historic narrative.

Nevertheless, Bostrom and Sandberg observe, we continue to approach these opportunities as if we have never left the caves. The various 'diseases of affluence' from which humanity suffers today – from ecological degradation to mediocre health maintenance and tolerance of massive socio-economic disparities – are traceable to a failure of our biological hardware to catch up with the aspirational software (i.e. the progressive philosophies, sciences, etc.) that successive generations of humans have programmed into the hardware. For Bostrom and Sandberg, then, 'the wisdom of nature' amounts to a euphemism for natural selection's *senility*, a solution to which is the transhumanist promise to upgrade humanity's hardware.

The proposal involves expediting our acquisition of the relevant competences – be it by gene therapy, nanobots or silicon implants – so as to remove our residual Palaeolithic tendencies in a matter of decades rather than having to wait for blind processes to act 'naturally' over aeons.

In short, whatever conception of 'evolution' transhumanists wish to support, it is unlikely that Darwin would have subscribed to it – even if he had had the benefit of today's science and technology. Darwin would have denied both of the following possible transhumanist relationships to evolution: (1) Humanity provides direction to an otherwise directionless evolutionary process. (2) Humanity fulfils evolution's latent potential. Both options should be familiar from the previous chapter as 'theomimetic' aspirations. Nevertheless, Bostrom and Sandberg cannot be blamed for wanting to see a seamless transition from Darwin to 'Humanity 2.0'. Darwin's own staunchest defenders have been diligently retro-fitting Darwin to enable him to carry the torch for the increasingly purposeful and efficacious ways in which we have been able to intervene in natural processes.

Consider the following remarkable passage, which is meant to be a damning review of Fodor and Piatelli-Palmarini (2010), a book that accused Neo-Darwinists of false advertising for claiming that 'natural selection' was non-teleological. Implied in their critique is that because Neo-Darwinists are so keen to distance themselves from creationists and other theists who claim that nature is subject to intelligent design, they have deliberately obscured their own dependence on the idea that 'nature' exhibits some sort of intentional structure. Needless to say, Neo-Darwinists refused to take this charge lying down. Consider the scorn poured on Fodor and Piatelli-Palmarini (2010) by the historian Robert Richards:

> The concepts 'selection for' and 'constraint on' do indeed have intentional properties, because they express judgments made not by nature but by human beings – intentional judgments by biologists to the effect that in particular environmental situations certain features of traits are causally relevant. Quite routinely, for example, medical experts attribute the evolution of drug-resistant strains of bacteria to the excessive use of antibiotics in hospitals– or in cattle feedlots. No hospital workers or cattlemen intend to select for drug-resistant bacteria, although their actual intentions

> obviously play a causal role. Scientists understand quite well how selection operates in these instances; indeed, they are able to breed drug-resistant bacteria experimentally precisely in the way these organisms are selected for in the 'wild,' thereby confirming the natural selection of drug resistance.
>
> Had Fodor and Piattelli-Palmarini read the first chapter of the *Origin*, they would have seen that Darwin argues there not so much that artificial selection is a model for natural selection as that it is exactly the same thing. Darwin regarded the breeder's intention, correctly I believe, as simply another environmental condition – one that rarely has a predictable outcome, as he discovered when he tried to breed fancy pigeons back to their original ancestral colors. Darwin thus directly demonstrated natural selection at work. And we do the same in the case of drug resistance. (Richards 2010)

Think about Richards' critique in light of our previous discussion. Stripped of his rhetoric of high dudgeon, Richards is simply saying that our failed attempts to turn nature to our advantage enables us to discover that nature has a mind of its own, from which we might learn so as to control it (experimentally) and thereby realize our original aim of turning nature to our advantage. The obvious question, then, is whether there is anything more to nature's 'mind' than a reification of our own ignorance of what prevents us from achieving our aims. This is a familiar German idealist trope for glossing 'nature', which in a more materialist mode corresponds to 'unrealized potential' or 'unexploited capital'. To be sure, this is exactly the opposite of how Darwinists wish to cast the issue. They prefer to say that learning more about nature's *modus operandi* amounts to discovering the objective limits within which we can effectively operate. Without outright personifying natural selection, Darwinists treat it as a force designed to keep our ambitions in check, a general principle of species domestication that includes *Homo sapiens*. Under the circumstances, it is difficult to see how transhumanism can gain much headway in its more expansive ambitions to master evolution. Our own view is that natural selection should be treated in the creative manner of engineers vis-à-vis gravity, which over the past 300 years, in the spirit of idealism, has served to minimize any sense that *Homo sapiens* was meant to live with feet planted on the Earth.

4 Eugenics as a productive development of evolutionary theory

Despite Darwin's own diminished view of humanity vis-à-vis other species in the cosmic order, historic contributors to Neo-Darwinism typically verged on positions that are now recognizably 'transhumanist'. Julian Huxley, the British founder of the Neo-Darwinian synthesis, coined the term 'transhumanism' precisely to capture the idea that *Homo sapiens*, albeit the product of billions of years of evolutionary history, nevertheless is so constituted that we can acquire an unprecedented second-order understanding of the entire evolutionary process, which in turn enables – if not obliges – us to direct its future course (Huxley 1953, 1957). In effect, Huxley claimed that *Homo sapiens* is a miraculous mutation, or what the geneticist Richard Goldschmidt at the time called a 'hopeful monster' (i.e. a mutation that results in the systemic re-organization of the organism). We put the point this way because it helps to explain why Huxley facilitated the translation and publication of the works of Pierre Teilhard de Chardin, the heretical Jesuit palaeontologist who was proscribed by the Pope from publishing in his lifetime. Teilhard de Chardin (1961) held precisely this view as part of an attempt to read the Transfiguration of Jesus into evolutionary history as the first moment when *Homo sapiens* recognized its god-like potential, the full realization of which would come at the end of time (the 'omega point'), when divine creation is finally brought to fruition. We discussed the biblical basis for this interpretation at the start of the previous chapter.

Huxley's concern with Teilhard, far from being aberrant, was shared by the person who published the most pedagogically influential formulation of the Neo-Darwinian synthesis, Theodosius Dobzhansky (1937). Indeed, when Dobzhansky (1973) famously told biology teachers that 'Nothing in biology makes sense except in light of evolution', he was referring to divine creation as itself a test case for evolution's scope, in terms of which – so he informed the teachers – Teilhard's work provided confirmation for his claim. Although Dobzhansky (1967) no less than Huxley held that the Neo-Darwinian synthesis could be defended on strictly scientific grounds, again like Huxley he believed that *Homo sapiens*' place in evolutionary history had substantial metaphysical import that pointed towards a species-wide moral imperative, whereby we take responsibility for the future of the cosmos. In the generation after the Second World War, this unashamedly anthropocentric, progressive view of evolution was still informed by the political movement that dared not speak its name, 'eugenics', as Huxley and Dobzhansky each served as president of his respective national (UK and US) eugenics society.

It is common among historians and philosophers of modern evolutionary theory to marginalize the cognitive significance of the theology and politics of the founding theorists, especially when these verged on the question of identifying or producing a superior version of *Homo sapiens*. These matters are either passed over in silence as an irrelevance or they are exoticized as the products of personal idiosyncrasy. Michael Ruse (1996, 1999) is the master of this strategy of obfuscation. The result is an artificially neutral presentation of the Neo-Darwinian synthesis, one that occludes the big fact that it is very difficult to find a leading scientific evolutionist who has not supported either (*laissez faire*-style) naturally selected genetic differences that, once correctly identified, might be understood as 'races' or (more proactively) the prospect of gaming such differences to humanity's overall advantage through planned reproduction policies (Pichot 2009). Differences of theology and politics, often quite subtle, are crucial for getting the full measure of this pronounced divide in evolutionary sensibilities.

In this context, an interesting contrast in background theologies is provided by Ronald Fisher and Sewall Wright, respectively, the UK and US founder of modern population genetics and both eugenicists. In particular, they offered rival interpretations of how natural selection could be understood in terms of changes in gene frequencies in an ecology. This was crucial to address critics of Darwin who regarded natural selection as merely a metaphysical principle capable only of rationalizing but not genuinely explaining evolution. Both Fisher and Wright were instrumental in translating natural selection into general mathematical formulae that could be used to model experimental situations that in turn simulate what normally transpires in nature. Although it is philosophically unfashionable to put the matter this way, both Fisher and Wright succeeded in injecting a 'Newtonian' sensibility into Darwin's original natural-historical approach. Transhumanists should prick up their ears at this point. We shall first discuss the sort of reading of Darwin that they jointly opposed – as should all transhumanists. Afterwards we turn to Fisher's and Wright's rather different but equally proactionary sensibilities.

Consider a long popular, late twentieth-century way of understanding Darwin's theory associated with the Harvard palaeontologist Stephen Jay Gould, which equates 'evolution' with the unique unfolding of life in geological time. Gould (1988) famously claimed that were evolution to happen all over again, it would most likely take a different course. Here he was being faithful to Darwin's own self-understanding, inferring a sense of fatalism related to the ultimate arbitrariness of life itself. However, Fisher and Wright saw the metaphysical implications of such contingency rather differently from Gould. They interpreted the supposedly arbitrary nature of our existence as simply the enactment of one among a very large number of possibilities, each of which could be realized under the right conditions. The implied parameters feature as variables in mathematical equations that enable us to acquire a general understanding of the 'possibility space' within which life flourishes (Frank 2011). To take a striking case in point, unlike Gould and today's Neo-Darwinists, Fisher was indifferent to the age of the Earth, since he understood what he dubbed 'The General Theory of Natural Selection' – something that might enjoy the universal validity of Newton's law of gravitational attraction – as a timeless principle whose plausibility did not depend on billions of years of chance-based processes working their magic (Ruse 1996: 295).

Put provocatively, Fisher and Wright were in search of God's room to manoeuvre (*Spielraum*), an aspiration that Gould thought, in classic Epicurean fashion, was either nonsensical or beyond human comprehension. In theological terms, Fisher and Wright are best understood as extreme Christian dissenters attempting to reverse-engineer God's strategy for optimizing matter's productivity. To be sure, Fisher and Wright dissented in opposing directions in their attempts to fathom this cosmic intelligence: Fisher maximized our distance from knowing the divine plan except in its most general form, while Wright minimized that distance by effectively rendering us co-creators. But both were centrally concerned with what in the previous chapter we called 'theomimesis', God-playing.

Fisher dissented in the direction of an *ultra-transcendent* deity, whose plan we can fathom obliquely in mathematical terms, such that we can identify and correct for deviations from the plan in the way we manage the human condition, without necessarily grasping the sort of world towards which 'the general theory of natural selection' is aiming. Fisher's brand of theology is a Neo-Calvinism that would have met with the approval of the original theologian of population pressure, Thomas Malthus, who also saw God's hand in the mathematics of what we now call 'carrying capacity'. On this basis, Fisher argued that human societies have impeded the workings of natural selection by allowing for the inheritance of wealth, which has resulted in bottlenecks in the distribution of opportunities for future generations. It is easy to see what Fisher means as occurring simultaneously at the level of economics and genetics, insofar as human legislation makes it possible for wealthy families to join together in marriage so as to perpetuate the amassed worth of their 'capital', understood in the broadest sense of the term. Against this backdrop, Fisher followed the founder of eugenics, Francis Galton, in regarding such concentrations of wealth as 'artificial', which is to say, detrimental to humanity's long-term survival. In particular, Fisher believed that the legal protection afforded to these scaled up family units encouraged a profligate lifestyle that contributed to the misery of those excluded from such biologically based corporate mergers (Box 1978).

In contrast, Wright dissented in the direction of an *ultra-immanent* deity, by which we mean that the divine plan requires humanity for its completion. In this respect (if perhaps in no other), he agreed with Teilhard de Chardin (1961) in regarding *Homo sapiens* as representing the leading edge of divine creativity, which Teilhard in particular believed might eventuate in the self-realization of God in matter through some ultimately advanced ('omega') version of ourselves, which may be glossed biblically as the return of Jesus. In the second half of the nineteenth century, a secular version of this position – that God is an emergent property of natural evolution – was effectively the house metaphysics of the German scientific community. Its most neutral name was 'monism' (Weir 2012). However, versions might stress either the realization of spiritual ideals in matter (e.g. the 'panpsychism' of Gustav Fechner, the founder of psychophysics) or the spiritualization of matter through an increase in organizational complexity (e.g. the 'hylozoism' of Ernst Haeckel, Darwin's German bulldog). Late in life, Wright (1977) made it clear that he belonged to the former group. In practice, he believed that we possess the spiritual competence to select from a population's offspring those mutations which, given a chance to increase its numbers in a protected environment for several generations, could then be released back into the larger population to expedite biological progress. An apt comparator for Wright's interventionism is Maxwell's Demon, a hypothetical being whose own work of directing the course of molecules enables a more efficient organization of matter than would otherwise naturally happen.

Seen in a positive theological light, we might regard both Fisher's and Wright's 'eugenic demons' as angels in disguise. Thus, while Fisher's angels are market regulators correcting misguided human attempts to interfere with God's intelligent design, Wright's angels are entrepreneurs capable of taking advantage of opportunities that nature throws up in order to bring the divine plan to fruition. But given the political controversies that would attend the application of either Fisher's or Wright's proactionary approach to genetics to humans, it is easy to see the rhetorical appeal of presenting the Neo-Darwinian synthesis is as normatively neutral terms as possible. For this, gun-shy biologists owe an enormous debt to philosophers of science who over the past half-century have constructed a semantic edifice known as the 'logical structure of modern evolutionary theory' that functions as a bulwark protecting the science from being sullied by its potential and actual applications. We already saw in the previous section how this bulwark operated to prevent anything like Hogben's socio-biological research programme from being revived after the Second World War, namely, by leveraging a multi-level analysis of evolutionary causation into a general scepticism about the prospect of efficacious eugenic interventions.

At one level, this defensive reaction by evolutionary biologists and their philosophical under-labourers is perfectly understandable. After all, anti-evolutionists – if not anti-science advocates more generally – continue to exploit the memory of twentieth-century eugenicist excesses across the entire scientifically developed world, not only Nazi Germany but also the US and Scandinavia, all informed by ideas of largely British origin. However, transhumanism cannot derive much justification for its aspirations from the sort of politically sanitized 'logical' account of evolution promoted by philosophers of science today. As Julian Huxley would have been the first to admit, much of what is nowadays proposed under the name of 'transhumanism' is simply 'Eugenics 2.0'. (The clear exception is the 'techno-gnosticism' of Ray Kurzweil [2005] that prophesies a long-term post-genetic future for humanity as our souls migrate from a carbon to a silicon platform.) Eugenics 2.0 is what eugenics looks like once interventions into the gene pool go beyond the gross regulation of individuals' breeding patterns to much more targeted interventions such as drug-based gene therapies and direct nano-level genetic re-engineering (Comfort 2012).

At this point, it is worth reflecting briefly on the normative stance adopted by Huxley himself as UNESCO's first scientific director vis-à-vis the Nuremberg Trials on the 'crimes against humanity' committed under the Nazi regime. Much to the consternation of moralistic historians (e.g. Weindling 2004), Huxley remained very much a eugenicist throughout the proceedings. He seemed to be relatively uninterested in whether scientists knew or approved of Nazi atrocities or, for that matter, the conditions under which human subjects participated in the most extreme forms of research. Rather, he was more interested in whether the Nazi research framework had been selected and applied in a scientific manner, as well as whether the suffering inflicted by Nazi research had resulted in significant cognitive benefit that might partly redeem its patent ethical deficiencies.

Huxley helped to salvage the careers of Nazi-friendly scientists such as the ethologist Konrad Lorenz (later a recipient of the 1973 Nobel Prize for medicine or physiology) who were not involved in human-based research. He also was the principal author of the landmark UNESCO 'Statement on Race', which for nearly two generations persuaded both natural and social scientists that race is a 'social construct' in a sense that implied its empirical invalidity (Brattain 2007). However, this creature of committee expressed a more politically correct but scientifically coarser view than that of Huxley himself. Despite the Nazi use of genetic differences to declare a race war that ultimately turned to genocide, Huxley was not deterred from believing in the likelihood of politically relevant genetic variations in the human condition. Huxley realized that once the fog of war has cleared, there remains a difference between accepting the existence of such variation and what society decides to do about it. The 'is' merely prompts but does not determine the 'ought'. Huxley is not given due credit for having stayed the course on this point.

An interesting point of comparison is Ernst Mayr, perhaps the most influential evolutionary taxonomist of the twentieth century, and someone who before immigrating to the US had been trained in the same tradition of medical science that bred the Nazi-friendly Lorenz. Even at the end of his very long career, which extended to the start of the current century, Mayr (2002) did not shy away from suggesting that health research and provision might be rationalized by paying increased attention to the genetic bases of disease. Taken as a scientific proposal, this seems quite reasonable, especially given that Mayr delivered it from a Harvard chair in a period of relative national wealth. However, it was just this sort of directive that throughout the twentieth century had played into the politics of restricted immigration during economic crises, fuelling xenophobia and racism internationally.

Consider the discipline of 'racial hygiene', which flourished in German medical schools in the first half of the twentieth century and rivalled sociology as the foundational social science discipline (Proctor 1988). To the racial hygienist, if certain medical liabilities are genetically marked, then depending on their prevalence in the society and the resources available, one might decide to incorporate them through domestic taxation (e.g. the unmarked subsidizing care for the marked as part of expanded national health coverage) or deport the relevant genetically liable individuals to countries better equipped to care for them, preferably with the full costs of relocation covered by the deporting country.

In short, the Neo-Darwinian synthesis constitutes a house divided against itself with regard to the emerging proactionary and precautionary world-views. In this section, we have highlighted the more proactionary genetic side, due to Mendel and more lab-based researchers specifically concerned with the transmission of traits from one generation to the next, adopted a more experimental spirit (e.g. inducing novel mutation, combination, segregation). As we have suggested, such leading twentieth-century geneticists as Fisher, Wright, Huxley and Dobzhansky intercalated their eugenicist interests and their scientific and theological views in rather different ways, but all with an aim of human self-improvement and empowerment. In contrast, the more precautionary natural history side of the synthesis, due to Darwin and more field-based researchers, focused on the ecologies where organisms with certain traits have enjoyed long-term survival. They tended more towards conserving the environments that have enabled life to become fit for purpose (or 'adaptive'), a quality that they believe is lost at our collective peril. In times of economic crisis, when *Spielraum* has appeared limited and the prospect of loss has loomed larger than gain, the latter Darwinian approach, often travelling under 'the wisdom of nature', has prevailed – which perhaps explains why transhumanists such as Bostrom and Sandberg wish to travel under that misleading banner.

Nevertheless, should there be any doubt in the reader's mind, either side of the Neo-Darwinian synthesis may be spun in a liberal or authoritarian direction. For this reason, we should not be too quick to make political judgements. Indeed, everyone in this debate saw themselves as occupying the 'Left' of the political spectrum, if we mean the party dedicated to using science to advance (or 'enhance') the human condition. An important difference that emerges from focusing on the precautionary/proactionary distinction is whether the advancement of the human condition ultimately requires the *conservation* or the *substitution* of the natural environment: precautionaries stress the former, proactionaries the latter.

Against this backdrop, the phrase 'National Socialism' should not be seen as either a sham or an oxymoron. A good reference point for this discussion is Franz Neumann's *Behemoth* (1944), a classic work on the political economy of Nazi Germany that is nowadays neglected because it was written in ignorance of the Holocaust and hence easily dismissed for having failed to see the 'true nature' of the Third Reich. Assumed in this judgement is that any savvy political observer in the early 1940s would have foreseen the Holocaust as the inevitable outcome of Nazism. Yet, Huxley, though certainly knowledgeable and outspoken against Nazi atrocities, did not treat the matter that way – and neither should today's transhumanists. The idea of the Holocaust as the inevitable outgrowth of modern science, and even the Enlightenment, a thesis that Zygmunt Bauman (1991) popularized to a generation of social theory students who found Adorno too difficult to read, suffers from the fallacy of 20/20 hindsight. However, Bauman's error, because it comes from someone who presents himself as a progressive thinker, poses a far greater threat to the pursuit of proactionary inquiry than, say, the pious 'higher yuckery' of a bioconservative such as Leon Kass (1997) or Francis Fukuyama (2002), as briefly discussed in Chapter 1.

A perversion of an adventurous scientific impulse – however heinous – should not obscure the overriding value of nurturing such an impulse. For example, Nazis took seriously the idea that in the interest of global sustainability, ecological management required that races should be encouraged to move to environments where their traits are most adaptive. To be sure, this line of thinking denied the sort of universalism jointly promoted at the time by capitalism and Marxism, both of which valorized our free mobility and behavioural plasticity. Here the work of Hayek's doctoral supervisor, Othmar Spann, provides the relevant alternative sense of 'universalism' as 'globalism': instead of presuming that everyone is capable of the same mode of existence, one presumes a diversity in modes of being that self-organize into a sustainable world-order (cf. Latour 2013).

The legacy of this line of thought continues to be felt in countries (notably in Scandinavia) that dedicate a relatively large portion of their budgets to overseas development aid, in the hope that the beneficiaries might develop their own welfare states customized to the specific health needs of a relatively restricted gene pool, just as the benefactors themselves had done. In that case, there would be no need for mass immigration to Northern Europe and its resulting socio-economic conflicts. On this basis, a leader of the German racial hygiene movement, Alfred Ploetz, an avowed pacifist who nevertheless ended his life as a Nazi sympathizer, was nominated for the Nobel Peace Prize in 1936. It is no accident that the maintenance of genetic equilibrium was a key custodial role of the Scandinavian welfare states from their inception, with radical deviation in the range of individual traits taken as either grounds for quarantine and elimination or cultivation and promotion – that is, the difference between defectives and geniuses – with most political interest focused on dealing with the defectives (Broberg and Roll-Hansen 2005).

Those who still wish to sharply divide Scandinavian sterilization policies (which inspired Hitler and continued in Sweden until the 1970s) from the Nazi genocide are themselves victims of moral essentialism, as if there were some absolute divide between good or bad in these matters. In fact, it is a slippery slope. After all, even the most politically liberal eugenicist – for example, today's supporters of 'pro-choice' abortion policies – is inclined to think of the prospect of an unwanted offspring as a threat to one's own existence (aka 'liberty') for a variety of reasons, which of course others who take more seriously the sanctity of 'potential life' may wish to judge differently. What the Nazis did was to scale up this personal concern into a nationwide problem, in which a very generalized sense of 'self' was mobilized by conjuring up the spectre of a potential mass attack from genetic invaders. If we just stick simply to the policy motivation (rather than the undoubted brutality of its implementation), it was neither based on 'bad science', in the sense of failing to rely on the best knowledge available, nor did it require the hijacking of 'good science' by bad people. Rather, very much like a Greek tragedy, it involved the gradual loss of proportion in a policy that might have worked in a more circumscribed application.

To drive the point home, if today's transhumanists still wish to fly the Darwinian flag, as the phrase 'the wisdom of nature' suggests, they need to take a view on the original – and peaceful – 'final solution' suggested by the Nazis, which was to divide the world into 'homelands', each adequately provisioned to deal with its natives' welfare. When first proposed it met with the approval of Zionists (Glad 2011). Of course, the default libertarianism of most transhumanists suggests that such a policy would be odious. Our increasing knowledge of the complexities of gene expression in the wake of the DNA revolution makes it easy to dismiss the proposal of homelands as appealing to an outdated genealogical populism, an earthbound version of the astrologer's appeal to stellar attractions. Nevertheless, it is just this strategy that continues to elicit a large volunteered scientific data pool – mainly in the form of individuals contributing their saliva for DNA scanning – that is providing a clearer picture of *Homo sapiens*' 'routes out of Africa' (Fuller 2006: chap. 8). All of this effort may result in a new set of genetic subdivisions of humanity that are more closely aligned with differences in researchable and treatable biomedical conditions. At least, that is the hope. In the meanwhile, there is evidence that the public is already mentally preparing itself for this prospect by reviving the social relevance of racial distinctions (Phelan et al. 2013). The question, then, from a bio-economic perspective, is whether the human phenotype has 'extended' sufficiently to accommodate the idea that all people should be allowed free movement in space – and, for that matter, free movement within their own bodily spaces, or 'morphological freedom' (Bostrom 2005).

Instead of demonizing the very idea of racial differences, we might consider it in the same spirit as the concept of division of labour, which is seen as both valuable and realistic even as the exact criteria for distinguishing jobs (or, by analogy, races) shift over time, given changes to both the economy and our understanding of human behaviour. Unfortunately, ever since Plato's 'myth of the metals' in the *Republic*, the West has tended to think of race in static, essentialist terms, thereby posing a 'natural' barrier to, say, one's capacity to be part of an ethnically heterogeneous nation-state (Hannaford 1996). However, a more scientifically informed, fluid notion of race might be part of a progressive sociology. In fact, if one considers the history of eugenics from a global perspective in the twentieth century – given that it was practised in some form virtually everywhere on the planet – it is exactly this fluidity that the field's advocates had in mind (Bashford and Levine 2010). Here it is especially interesting to examine Latin American contributions to discussions of ecology and eugenics.

In the case of ecology, we specifically mean the political economy of the Incan Empire, which the Spanish conquistadors discovered to have been based on a vast, intricate understanding of what we would now call 'biodiversity' (Cañizares-Esguerra 2006: chap. 6). Indeed, this 'equatorial paradise' inspired Carolus Linnaeus two centuries later to conclude that humanity was capable of manufacturing Eden-like conditions throughout the Earth, literal 'microcosms' that would contain life forms in their full variety. The legacy of this idea remains in the modern institutions of zoological and botanical gardens, albeit not quite as the 'meta-level farms', so to speak, that Linnaeus himself had intended. Their spiritual descendants are better located in the agricultural stations that served as laboratories for Fisher and Wright (Koerner 1999).

In the case of eugenics, we mean the brilliant counter-European move championed by Mexico's answer to Wilhelm von Humboldt, José Vasconcelos, the first rector of the Autonomous National University of Mexico (UNAM), who placed a positive value on people being of mixed race, or 'mestizoism' (Stepan 1991: chap. 5). The idea here is that 'constructive miscegenation' amounts to a dialectical synthesis of humanity's full potential, possibly resulting in feats that would not have been achieved by pure race individuals. To be sure, outside of Mexican nationalist contexts, Vasconcelos' vision was easily dismissed as relying on what was by then (the 1920s) an empirically discredited Lamarckian understanding of evolution, whereby these purpose-bred mestizos would somehow transmit their progressive world-view to future generations (Miller 2004). Nevertheless, the idea of Mexico as the breeding ground for what Vasconcelos called the 'cosmic race' – a radicalization of the US self-understanding as a 'melting pot' (which had been *de facto* only of European races) – animated the internationalist imagination prior to the onset of the Cold War, not least the artist Diego Rivera and his friend, Leon Trotsky, who spent his final years exiled in Mexico City. The legacy remains in UNAM's motto: '*Por mi raza hablará el espíritu*' ('The spirit shall speak for my race'). In the context of transhumanism, the motto is best understood as the ambition to produce people capable of embodying our full humanity, a task that requires *both* education and eugenics.

4
A Legal and Political Framework for the Proactionary Principle

1 The current legal standing of the precautionary and proactionary principles

An important obstacle prevents the proactionary principle from being accorded the same legal standing as the precautionary principle. Whereas the proactionary principle was clearly meant as an equal and opposing imperative to the precautionary principle, in practice it tends to be treated as a modification or attenuation of the precautionary principle. Lost in the translation is that the two principles project alternative universal ideals for the human condition that are not obviously compatible with each other. Both principles are, in philosophical parlance, 'axiological'. They are concerned with values, specifically: *what is a human being for?* The precautionary says that we are part of a larger whole called 'nature' and the meaningfulness of our lives (not to mention our sheer survival) is based on our appreciating that deep metaphysical point. It is all about *self-embedding.* In contrast, the proactionary says that we are no mere part of nature; rather our existence gives meaning to an otherwise meaningless nature by serving as means to our ends. The precautionary wishes to return us to our biological origins, the proactionary to take us as far away from them as possible through endless acts of *self-transcendence.* This difference is ultimately grounded in the value of being human: are we animals suffering from too big brains (precautionary) or deities in need of more resources (proactionary)?

This helps to explain why replacing the natural with the artificial is so key to proactionary strategy, and why some proactionaries speak nowadays of 'black sky thinking' that would have us concede – at least as a serious possibility if not a likelihood – the long-term environmental degradation of the Earth and begin to focus our attention on space colonization (Wilsdon and Mean 2004). To the proactionary, our biological inheritance, however long in its evolution, presents us with the ultimate challenge to our creative powers: If we are the gods that we think we are, then we need to demonstrate our capacity to transcend our current material basis – even our carbon basis, if Ray Kurzweil (2005) is to be believed. In Chapter 1, we characterized the Proactionary–Precautionary binary as a 90-degree rotation on the classic Left–Right ideological axis.

However, this sense of polarity has been so far lost. Perhaps the most sophisticated version of the problem is in evidence at the European Commission (EC), which treats the precautionary principle as a normative anchor for European Union legislation. René von Schomberg, student of Jürgen Habermas and resident philosopher at the EC's directorate for research, has penned many works in defence of 'responsible innovation' as a general policy that is underwritten by the precautionary principle (e.g. von Schomberg 2006, 2013). While von Schomberg intends this policy as a spur to innovation in exactly the areas most central to the transhumanist agenda – nano-, bio- and info- sciences and technology – he couches it in terms of the precautionary principle for two main reasons: (1) its standing in international environmental law and (2) its conformity to the conventional understanding of welfare-state action as aiming to protect (rather than promote) people. Taken together, these two reasons create a presumption that we should always worry about who will be harmed before who will benefit – regardless of the exact nature and number involved in both cases.

While superficially a humane policy, the privileging of harm over benefit is 'humane' *only if you believe that, above all else, disruption to your default way of being is the worst thing that can happen to you.* We put the point in italics because proactionaries envisage that there are still worse things that could happen to you: in particular, you might be (paternalistically) shielded from opportunities that would have afforded you – albeit perhaps via some risky personal behaviour – a substantially better state of being. This is the spirit in which we strongly support moves towards a 'duty' or 'right' to science. In any case, a clear implication of this discussion is that the proactionary principle cannot be understood as simply a modification or attenuation of the precautionary principle. Rather, the two principles are fundamentally opposed in terms of the relative priority of harm and benefit in their respective welfare functions: proactionaries weigh benefit over harm, precautionaries harm over benefit. Moreover, as the case of von Schomberg illustrates, support for the precautionary position is often tied less to some overriding Epicurean metaphysical commitment to avoid pain whenever possible than to a certain conventional interpretation of how to address issues of 'responsibility' in cases of (political or economic) agents whose actions have the capacity to inflict pain on others.

In contrast, according the proactionary principle the same legal status as the precautionary principle would involve treating it as a universal prescription that sets the burden of proof on those who would prioritize protecting over promoting the human condition as an aim of the law. In welfare economics, one speaks of this difference in terms of 'lexicographic ordering': we all want to maximize benefit and minimize harm, but which imperative takes precedent if we cannot pursue both as much as we would like? To be sure, the question is more complex than it seems: are we talking about benefit/harm across a society's members at a particular time or over a long expanse of time? For example, precautionaries might be happy to avoid long-term harm even if that does little more than enforce a sustainable version of our current condition, whereas proactionaries might well seek large long-term benefits for survivors of a revolutionary regime that would permit many harms along the way. However, in practice, the choices are rarely so stark. As the late Ronald Dworkin (1977) made a career of reminding us in the context of US Constitutional law, the role of judges is to make a justifiable trade-off between countervailing principles to resolve particular cases. However, such trade-offs do not invalidate the ideals represented by opposing principles; on the contrary, they provide springboards for larger political discussions of the underlying values represented by those ideals.

Thus, a reformed European Commission might produce policies that sometimes veer in a precautionary and other times a proactionary direction, but each case would afford an opportunity for thinking about the diametrically opposed visions of the future of humanity that they project. Indeed, political parties originally formed in the late eighteenth century precisely to keep those conversations going even in the absence of relevant cases for legislation or adjudication. In a world now saturated with public relations, we might think of party loyalty as akin to brand loyalty, whereby adherence to the party or brand becomes a matter of unconditional commitment, regardless of whatever short-term losses might be incurred in the process. Thus, we have stressed the *ideological* character of the divide separating 'proactionary' and 'precautionary' as principles. However, there is also a more strictly legal angle to the divide – indeed, pointing to alternative ways of regulating the disposition of legally protected rights, or 'entitlements'. It comes from a classic article in the law and economics literature that proposes two universal ways of determining entitlements: by *property rule* or by *liability rule* (Calabresi and Melamed 1972). The precautionary principle privileges a property rule approach, the proactionary principle a liability rule approach.

In a property rule approach, an entitlement's bearer is clear but its value is negotiated in court; in the liability rule approach, the value of the entitlement is clear but its bearer is negotiated in court. The former favours the actual owners of capital, the latter its potential users: *rentier versus bourgeois*, as both Ricardo and Marx would have recognized. The property rule approach operates in the spirit that you don't shoot until you have received permission, whereas the liability rule approach permits you to shoot first and ask questions – that is, the terms of compensation – later. In terms introduced in Chapter 1, the former performs a 'politics of position', the latter a 'politics of momentum'. Perhaps the most widespread use of the liability rule approach occurs in the interpretation of copyright law with regard to the repeat playing of music (Lessig 2001: chap. 7). In this context, to minimize the transaction costs of every radio station having to seek permission to play a song, a condition of the granting of copyright is that its owner agrees in advance to a fixed compensation for each play.

We proactionaries look forward to the day when something similar might apply to those in possession of a unique genetic sequence whose regular manufacture (i.e. repeat performances) is necessary for various drugs and treatments. Of course, genetic identity copyrights presuppose a regime of genomic registering that has yet to exist but is likely to do so in the future. But such a policy is unlikely to empower individuals, unless insisted upon. At the moment, biometric data is usually sought by states for reasons of domestic security, specifically crime and immigration purposes. But when turned into a for-profit business, the result is 'bioprospecting', which is often conducted by organizations legally located outside the jurisdiction of the places where the relevant bio-matter samples are gathered. International legislation is required to ensure the fair transaction of genomic materials (Safrin 2004). Indeed, it may serve as a vehicle for redistributing global wealth, in the spirit of the so-called Tobin Tax on financial transactions. Moreover, the liability rule approach helps to dispel the mystification of ownership surrounding 'nature', which has been traditionally invoked to prevent self-alienation (Brown 2003: chap. 8). One of the virtues of capitalism – which most forms of socialism carry over – is that no special value is accorded to how things have been done in the past. Rather, what matters is whether the change has been adequately justified, which is to say, rewarded or compensated. Without pretending that this is easy to determine in particular cases, not least because of the moral and technical difficulties involved in the assessment of rewards and compensation, nevertheless the proactionary principle implies that this would be a good general direction for the law to take (Krier and Schwab 1995).

2 The proactionary vision of science as the moral equivalent of war

In 1906 the great American pragmatist philosopher William James delivered a public lecture entitled, 'The Moral Equivalent of War'. James imagined a point in the foreseeable future when states would rationally decide against military options to resolve their differences. While he welcomed this prospect, he also believed that the abolition of warfare would remove an important pretext for people to think beyond their own individual survival and towards some greater end, perhaps one that others in the future might end up enjoying more fully. *What then might replace war's altruistic side?* The very need to ask this question means that we no longer live in a world dominated by conservative ideologies based on 'heredity' or 'tradition', to invoke two terms that by default were capable of legitimizing quite specific patterns of altruistic behaviour. In such cases, one's identity was tied to occupying a well-defined place in the reproduction of the social order: *I serve, therefore I am.* In contrast, James's search for a ground for altruism occurs in light of two anti-conservative ideologies: *liberalism* and *republicanism* – and, at a more general level, *economics* and *politics* as disciplines (Fuller 2000: chap. 1). In both cases, the sense of relevant 'others' to whom one might be altruistic is not given by, say, 'natural law' but is a feat of individual or collective self-legislation.

The two anti-conservative ideologies may be analytically distinguished as follows: in each binary, the former term ('liberalism' and 'economics') refers to how the parts common to a whole relate to each other, the latter ('republicanism' and 'politics') to how the whole relates to whatever lies beyond its self-defined boundaries. (Those already living in such a whole tend to replace 'parts' and 'whole' with 'individuals' and 'society'.) In terms of political economy, the republican is concerned with the structural features that need to be reproduced for the polity to retain its identity over time and the liberal with the fair exchange of goods and services among individuals at a particular time that enables this larger goal to be realized. Thus, together the two ideologies presume that neither the polity's boundaries nor its internal relations can be determined simply on the basis of repeating past practice. Beyond the obvious point that past practice can never be literally repeated, their anti-conservatism is rooted in a refusal to presume that (*ceteris paribus*) past practice should be carried over into the future, simply by virtue of its survival. It is just this historical orientation that has provided the Left-ward tilt to modernity and explains the typically disdainful and residual manner in which classical sociology treated 'traditional' modes of authority.

It is telling that the most famous political speech to adopt James's title was US President Jimmy Carter's 1977 call for national energy independence in response to the Arab oil embargo. Carter characterized the battle ahead as really about America's own ignorance and complacency rather than some Middle Eastern foe. While Carter's critics pounced on his trademark moralism, they should have looked instead to his training in nuclear physics. Historically speaking, nothing can beat a science-led agenda to inspire a long-term, focused shift in a population's default behaviours. Louis Pasteur first exploited this point by declaring war on the germs that he had shown lay behind not only human and animal disease but also France's failing wine and silk industries. Nearly 150 years later, Richard Nixon's 'war on cancer', first declared in 1971, continues to be prosecuted on the terrain of genomic medicine, even though arguably a much greater impact on the human condition could have been achieved by equipping the ongoing 'war on poverty' with comparable resources and resoluteness.

That science in the service of war can focus the collective mind as nothing else can was first realized in the Franco-Prussian War of 1870, in which Pasteur and Robert Koch were pitted as opposing germ warriors fighting to protect their nations' respective troops from succumbing to anthrax in the battlefield. But this insight truly came into its own only a half-century later, in the wake of the First World War. Partisans on all sides observed that the impressive consolidation of resources that ensured the health, safety and progress of the armed forces could be used as a template for promoting social welfare in peacetime. Here the Marxist sociologist and logical positivist Otto Neurath sang in harmony with the liberal economist John Maynard Keynes, both urging what Keynes called 'the great experiment' that would have the wartime economy carried over into peacetime by allowing the state to sustain a very tense production schedule that simultaneously stresses risk-taking, comprehensiveness and efficiency (Proctor 1991: chap. 9). In practice, it would mean a faster pace of invention and replacement, as the populace remained focused on everlasting foes – indeed, of the sort that William Beveridge identified at the end of the Second World War as the 'five giant evils' (want, disease, ignorance, squalor, idleness) against which the British welfare state would be designed to stand guard.

An historical argument that weighs in favour of the tight productive link between science and the military cast of mind is that World War III never happened. The feared nuclear Armageddon, imagined to be exponentially deadlier than the first two world wars, was for nearly a half-century sublimated as a 'science race' before economic burdens on both the American and Soviet sides (but especially the latter) finally extinguished any desire for open conflict. While this 'Cold War' is now easily dismissed as a black hole for scientific talent that was fixated on producing both ever smarter weapons and ever smarter means of defeating them, we continue to dine on its by-products. The most obvious case in point is the internet, which began life as an internal network for researchers funded by the US Defence Department who might need to pool their data on very short notice in case of national emergency – but, of course, also in the absence of any such emergency. This was the context in which the internet's full potential was realized in peacetime and which over the past quarter-century has revolutionized the very conduct of human life.

On this basis the US political scientist Daniel Sarewitz (2011) has urged that policymakers otherwise inclined to extend 'market-based solutions' to research funding should not overlook the unique combination of foresight and experimentalism historically displayed by the US Defence Department's 'war economy' model. In many fields, the military was a majority shareholder in research policy during the Cold War. However, instead of acting complacently, as one might expect of a majority shareholder, those on military contracts were inspired to unprecedented feats of innovation by virtue of having to respond to a comprehensive yet unpredictable external foe. Might the prospect of ecological collapse, epidemics or even global financial meltdown not serve a similar function to focus minds in our own day? If so, then a 'moral equivalent to war' will have been certainly found, though the exact duties and responsibilities that this state of conflict requires remain to be addressed.

Science's ability to step in as war's moral equivalent has less to do with whatever personal authority scientists command than with the universal scope of scientific knowledge claims. Even though today's science is destined to be superseded, its import potentially bears on everyone's life. Science functions as a kind of 'meta-politics' in democracies, which can intervene in the normal political process in a manner that is normally reserved for attacks from an external foe. From the standpoint of the history of science, the atom – the ultimate unit of matter – was a metaphysical notion associated with ancient Epicureanism until the early twentieth century, when it became a foundational concept in physics. But from the standpoint of the history of politics, the prospect that some substance – atoms, as it turns out – might contain the source of all energy in the universe has fuelled the imaginations of everyone who has sought to take control of their fate from the medieval alchemists onward.

Unsurprisingly, then, the atom came to be seen as holding the secrets to national security, from warding off common enemies (i.e. nuclear deterrence) to securing personal freedom (i.e. energy independence). In the heyday of the welfare state, a vague but popular belief in this tight bond between the metaphysical status and the empirical impact of atoms licensed a 'multiplier effect' approach to the public finance of 'blue skies' science, whereby everyone would agree to have their incomes taxed to subsidize the training of those who, by virtue of their competence, are in the best position to exploit the opportunities afforded by research that in the long run will benefit everyone.

Proactionaries take this argument to the next level: people should agree not only to be taxed but also to participate personally in cutting-edge scientific research. In republican ideology, it would be recognized as an extension of national service, a duty of citizenship. Even in the heyday of the welfare state, that point was generally understood. Thus, in *The Gift Relationship*, perhaps the most influential work in British social policy of the past 50 years, Richard Titmuss (1970) argued, by analogy with voluntary blood donation, that citizens have a duty to participate as research subjects, but not because of the unlikely event that they might directly benefit from their particular experiment. Rather, citizens should participate because they would have already benefited from experiments involving their fellow citizens and will continue to benefit similarly in the future (Reubi 2012). In effect, their participation helps to repay the debt that one normally incurs simply from having been born in the womb and borne on the shoulders of others.

In this context, a remark about Nietzsche's *Genealogy of Morals* (1887) is in order. We accept his etymological demystification of obligation ('ought') as having descended from the discourse of debt, whereby the threat of punishment lurks behind any felt obligation. However, we draw more prosaic conclusions. Simply by virtue of being allowed to live, you are invested with capital on which a return is expected. The 'fear unto death' that the Existentialists took to be the leitmotif of our lives refers not to mortality *per se* but the prospect that you may not have lived up to your potential (i.e. the capital invested in you) by the time you die: you will not have provided sufficient return on (genetic, educative, etc.) investment. You failed to redeem your debt. While it is certainly true that you were not consulted before you were born, you cannot credibly question your state of being unless you draw on what you have been already granted, not least in order to establish your identity as someone who requires that others give you more than you currently have. In short: why should we believe that fate has dealt you a bad hand unless you are already holding a particular set of cards? In short: *I protest, therefore I am capitalized.*

Of course, some styles of 'protesting' are more effective than others in redeeming one's debts but in the end they are no more than responses to already incurred obligations. Christians and capitalists are united in believing that the burden of proof is always on those who would commit suicide in response to their existential plight. The despair of such individuals may be no more than a narcissistic response to their failure to understand the full range of their capacities and prospects – perhaps because political correctness prevents an open discussion of such matters. The proactionary solution to this problem is that children in their school years be already made aware of the various scenarios involved in their trying to achieve a range of life goals, in light of their particular scores, etc. For example, someone who consistently scores at the mediocre range in quantitative ability but wishes to become a mathematician needs to know the paths that are available – and the effort required – to achieve that goal. They might even be told which political programmes of 'affirmative action' or 'positive discrimination' are likely to expedite or impede that achievement, since political knowledge is no less 'epistemic' than the mastery of a skill or a body of texts. But what *cannot* be allowed in a proactionary society that claims to respect life is a lazy libertarianism that would license suicide simply based on an untutored introspective judgement of how the world responds to one's felt wants and needs. Individuals cannot be 'self-determining' in any sense that Kant would recognize unless they want to know the exact range of their capacities.

Unfortunately, any neat fit between science and altruism has been undermined over the past quarter-century on two main fronts. One stems from the legacy of Nazi Germany, where the duty to participate in research was turned into a vehicle to punish undesirables by studying their behaviour under various 'extreme conditions'. Indicative of the horrific nature of this research is that even today few are willing to discuss any scientifically interesting results that might have come from it. Indeed, the pendulum has swung to the opposite extreme. Nowadays elaborate research ethics codes enforced by professional scientific bodies and university 'institutional review boards' protect both scientist and subject in ways that arguably discourage either from having much to do with the other. Even defenders of today's ethical guidelines generally concede that had such codes been in place over the past two centuries, science would have progressed at a much slower pace (Schrag 2010).

The other and more current challenge to the idea that citizens have a duty to participate in research comes from the increasing privatization of science. If a state today were to require citizen participation in drug trials, as it might jury duty or military service, the most likely beneficiary would not be the general public, but a transnational pharmaceutical firm capable of quickly exploiting the findings for profitable products. In that case, what may be needed is not a duty but a *right* to participate in science. This proposal, advanced by Sarah Chan at the University of Manchester's Institute for Bioethics, looks like a slight shift in legal language (Chan et al. 2011). But it is the difference between science appearing as an *obligation* and an *opportunity* for the ordinary citizen. In the latter case, one is not simply waiting for scientists to invite willing subjects. On the contrary, potential subjects are outright encouraged to organize themselves and lobby the research community with their specific concerns. The cost of this proposal is that scientists may no longer exert final control over their research agenda, but the benefit is that they can be assured of steady public support for their work. Our 'hedgenetics' proposal, discussed in the next section, provides a legal context for fleshing out Chan's proposal of a 'right to science'.

But before turning to the legal basis for hedgenetics, it is worth underscoring that the lingering Nazi stigma around a 'duty to science' combined with the neo-liberalization of the science policy environment has delivered a one-two punch to the republican ideology that would most naturally license scientific inquiry as 'the moral equivalent of war' in today's world. To be sure, our 'hedgenetics' proposal is designed to revive that ideological sensibility in the guise of a 'right to science'. Nevertheless, it is instructive to start from a position where there are only free individuals who need to be persuaded that they should engage in the collective risky projects that are characteristic of the most interesting forms of scientific research. In other words, imagine a world in which liberalism has been set adrift from a republican normative sensibility, such that the radically self-interested ethic of libertarianism is given full sway. How might we get such individuals to recognize scientific inquiry as the moral equivalent of war?

A useful way to approach the question (suggested to us by Tomas Hellström) is as an instance of an 'inverse tragedy of the commons'. The original tragedy of commons was about the prospect that if everyone who shares a common resource derives the most short-term personal benefit from it, the overall long-term effect will be to deplete the resource so that no one can any longer benefit from it. The problem then is how to motivate people to exercise self-restraint in the name of the common good – if possible, short of outright coercion. In the ecologist Garrett Hardin's (1968) original formulation, now seen as a landmark in contemporary precautionary thinking, the only effective solution appeared to be the (benevolent) coercion of the state. A concrete proposal in this spirit is the call by a think-tank affiliated with the UK Green Party for an oversight parliamentary chamber to look after the interests of 'future generations' (Read 2012). In contrast, proactionaries face the inverse tragic situation: participating in scientific research is in the short-term quite risky for oneself but there are potentially quite large long-term benefits if everyone participates. A political version of the same quandary would be how to motivate individuals to participate in the overthrow of an unjust regime.

While we do not expect libertarians to be especially moved by the debt-based arguments that have grounded a 'duty to science' in the past, they might be moved by more Cold War-style arguments that point a common external foe, such as disease and despoliation, that cannot be effectively tackled by individuals acting alone on their own behalf – or, for that matter, by sticking with the status quo. Max More's (2005) original 'Proactionary Principle' manifesto is especially clear on this last point. In short, 'nature as enemy' might revive a republican sensibility even in the most self-serving libertarian heart. Here it is worth recalling that a classic formulation of humanity's transition from barbarism to civilization – common to Bacon, Hobbes, Comte and Spencer – is that people decide that it is in their own self-interest to replace fighting among themselves with fighting together against the their common natural limitations. This narrative reappears throughout modernity from explaining the rationale for the original social contract to the rise of industrial society, where competition in the marketplace sublimates the urge to go to war.

One specific science-based example of the entanglement of all our fates pertains to the management of the human gene pool itself.

As we saw in the previous chapter, eugenicists have urged that this sense of stewardship be built into the fabric of the welfare state, and clearly this informed Titmuss's considerations above. We shall devote the final section of this chapter to a legal grounding that enables people to take collective responsibility for their genome by virtue of sharing salient genetic patterns, something that we propose as a cornerstone of political and scientific identity in a proactionary welfare state. But before embarking on that task, let us first outline a general strategy to motivate individuals to adopt a more proactionary approach to welfare:

1. Teach people about the risks they already take (e.g. in their normal diet) to provide a baseline for comparing the risks in, say, participating in scientific experiments. (The philosopher of science Larry Laudan pioneered this approach in a set of popular books and articles in the 1990s, though it seems to have gained little traction.)
2. Show how people already benefit from the risks that others have taken – often on their behalf, albeit in some abstractly specified way (e.g. in war or in research – the Titmuss point).
3. Argue that it is open to everyone to steer the research agenda to their potential benefit by investing both their economic and biological capital (e.g. in the manner of 'hedgenetics' discussed below) in any of wide variety of research projects in which they might be able to demonstrate a personal stake.
4. Remove criminal sanctions from the conduct of risky experiments on the condition that all of the data gathered from them – especially negative outcomes – are publicly reported and freely accessible.
5. Impose higher taxes on those who remain reluctant to take the relevant risks even after having been informed of (1), (2), (3) and (4) in order to subsidize the risks taken by others, a new form of social insurance that pays out compensation in case of harm (but does not pay the risk-takers anything upfront).

3 'Hedgenetics' as an example of a proactionary socio-legal regime

Ever since Johannes Gutenberg invented the printing press in the mid-fifteenth century, links have been made between the communication of information and human dignity, progress and equality. Indeed, the capacity to know has shifted from being a status marker to an instrument for economic development and social inclusion. For the past quarter-century the United Nations has been in the forefront of this shift, originally a vehicle to prevent and control the spread of HIV-AIDS (United Nations General Assembly 1987). By the 2003 Geneva World Summit on the Information Society, the UN explicitly recognized the continuity between the classic human rights to free expression and free assembly that had been forged in the Gutenberg Era and the emergent claims for a universal right to internet access. By 2011, the Special Rapporteur to the UN Human Rights Council, Frank La Rue had gone so far as to recommend that 'the right to internet access' be accorded special second-order status as an 'enabler' of other human rights, thereby putting on pressure on states not to block the flow of information, even (as, say, China avowed) in the name of national security. Unsurprisingly the General Assembly took this line of reasoning to its logical conclusion, namely, a declaration that included human rights itself as something about which people had the right to access information (United Nations General Assembly 2011).

However, at the same time the reverse process of containing information, ideas and inventions and by attributing them to a particular (physical or legal) person continues apace. A watershed moment occurred in the late eighteenth century when the US Congress passed the first Patent Act enabling what we recognize today as intellectual rights protection. Unlike the early modern days when socio-economic position was expressed by the number of tomes in one's personal library, today we have come to accept the equation of 'knowledge = power' as straightforward proposition (Stehr 1994). We have come to believe that ideas, invention and information, being a product of one's thought, ought to be treated as a tangible asset equal to, if not more valuable than, the product of manual labour.

Moreover, as modern law broadened its cognitive horizons, so did science. In the second half of the twentieth century, the frame of reference for identifying human abilities and identity moved from the external to internal domain, centring on our genetic make-up. Humans, as never before, could see that they were constituted as not merely a thicket of blood vessels and a bag of bones but as a source of information about themselves, a bastion of biological identity that connected with past generations. However, this newly acquired source of identity faced two challenges: (1) misuse of patent laws and (2) developments of genetics that started delivering ambiguous data about what we can know about ourselves, what constitutes information. Together they effectively challenged the notion of what constitutes an invention. When law meets genetic ambiguity, notions of ownership, value and identity stand on shaky ground. The patentability of genes presupposes our capacity to dispose of them freely, yet this capacity rests on how we define our social and individual identity.

In what follows, we consider the patentability of genes and the related question of whether developments in genetics have increased or decreased our rights to self-determination. We endorse the idea of 'genetic stewardship', whereby individuals are entitled to cultivate the genetic potential of humanity by virtue of being its natural bearers. Our inquiry culminates in a proposal for a new entrepreneurial branch of applied science: *hedgenetics.* In this context, we discuss how gene ownership can be empowering to individuals and enhancing of a sense of group identity.

The patentability of genes is a highly polarized issue. However, the ensuing controversies have been less to do with how genome ownership might best contribute to the collective development of humanity than two other issues: (1) whether the isolation of genetic information constitutes merely a discovery of a naturally occurring process (and hence non-patentable) or, such activity, due to the apparatus and interpretation involved in 'reading' genetic material, should be classified as an invention (and hence eligible for a patent); (2) assuming the latter, who is entitled to benefits of genome ownership – the human sources of information, the scientists who extract and interpret genetic information, or the companies who finance genetic research? Our own view is that the question raised (1) reflects an obsolete understanding of science's relationship to the world, while (2), though important, is much too narrowly gauged in terms of competing interests.

In the recent landmark US Supreme Court decision of *Association for Molecular Pathology v. Myriad Genetics* (2013), the sheer existence of a specific sequence of DNA prior to its extraction and testing (in this case, for cancer) cancelled any patent claims by the Utah biotech firm Myriad Genetics, even though the process of isolating the specific genetic information as relevant to the human condition would seem to imply that the scientists concerned are providing some kind of 'added value' to nature. However, the Justices, in a unanimous verdict, focused entirely on the material pre-existence of the DNA, without taking into account the difference made by isolating a bit for a specific purpose that nature itself would not have done, left to its own devices. Fortunately, from the standpoint of future genetics research, the judgement did not extend to the synthetic transformation of existing genes. A relevant UK precedent is the House of Lords' landmark decision in *Kirin-Amgen Inc. v Hoechst Marion Roussel Ltd* (2004), which also made references to the 'natural world' as grounds for prohibiting patenting parts of the genome that may have been produced in some 'purified' form but not, strictly speaking, 'improved' from its naturally occurring form.

We hold that the overriding significance attached to such distinctions as 'natural' versus 'artificial', or 'discovery' and 'invention', in intellectual property law is thoroughly misplaced, largely because such distinctions are ultimately based on a scientifically outmoded metaphysics. However, this point is easily obscured because this branch of the law is laced with 'morality provisions' that permit special (often Green) interest group considerations to steer the direction of judicial opinion (Seville 2009: 104–41). This point has become increasingly apparent over the past 30 years, including such landmark cases as HARVARD/ONCOmouse (1991) and HOWARD FLOREY/ Relaxin (1995). In the case of *Myriad Genetics*, even nominally dispassionate analyses of the case offered by historians (e.g. Kevles 2013) seem to have been unduly coloured by the case's financial dimension. Specifically, Myriad Genetics had patented a breast cancer predisposition gene (BRCA), thereby becoming the only company that can sell tests for the presence of said gene. This effective monopoly over a sphere of human anxiety provoked outrage in those forced to pay $3000 to $4000 to get tested (Jacobson 2011).

To be sure, such outrage would seem to have historical precedent, certainly in the US. Consider the case of Henrietta Laks, a terminal cancer patient whose cells were harvested in 1951 without her consent and then commercialized (as 'HeLa' cells) for use in cancer research, after having become the first successful instance of human cell cloning. In the early 1970s, after the Laks family began receiving requests for blood samples from medical researchers to perform genetic tests, they realized that they – and other patients – were routinely undergoing procedures that contributed to the production of drugs for which they would be forced to pay, even though the drugs that would not have existed without their participation, either as blood donors or experimental subjects (Washington 2011: 28–44). Although the Laks case was settled out of court, a similar one that was brought to trial is *Moore v Regents* (1990), in which the court ruled that another cancer sufferer, John Moore, had no entitlement to the proceeds from the commercialization of his discarded body tissue by his attending physician, a researcher at the UCLA Medical Center. The judges ruled that Moore had no natural property rights over his body, and were he endowed with such rights, the future of medical research could be put at risk. At most, Moore was entitled to claim for damages, since his physician failed to tell Moore about his financial interests (but given his medical condition, it is unlikely that the additional knowledge would have caused Moore to refuse treatment). The apparently precarious proprietary hold that people have over their own bodies that was revealed in the *Moore* judgement ended up serving as the inspiration for Michael Crichton's (2006) *Next*.

In light of these precedents, the judgment against Myriad Genetics might be understood as a rebuke to the distribution of costs and benefits practised by pharmaceutical companies and other biomedical firms, but not a measured reflection on the patentability of genes. Nevertheless, in the US the courts are not alone in their attempt to challenge the patentability of genes. The Obama administration has cast corporations keen to patent genes as modern day slavemasters (Ledford 2010). However, this policy goes against Obama's Material Genome Initiative, which aims to double the speed with which America develops new materials by re-engineering the molecular bases of various substances in the name of improved health, clean energy and national security (National Science and Technology Council 2011). Since it is nearly impossible for companies to develop drugs without knowing who owns the relevant genes, companies are unlikely to expedite production if they experience legal ambiguity. In this respect, it is striking that American courts, after having allowed genetic patents for over 30 years, making it possible to patent certain living organisms, genetic procedures and proteins, are now reverting to a situation – post-*Myriad Genetics* – where genes cannot be patented at all. Nevertheless, the pressure on genetic research to keep expanding remains. Since the landmark case of *Diamond v Chakrabarty* (1980), which allowed genetic patents, 3000–5000 American patents on the human genome and 47,000 patents on genetic interventions have been registered (Cook-Deegan 2009: 69).

It is fair to say that first genetic patents exploited the immature state of this aspect of patent administration as well as the relatively low level of scientific knowledge about genes. Nevertheless, it appears radical to overturn the law in this regard. When *Diamond v Chakrabarty* was decided, American courts were dealing with an unprecedented number of intellectual property cases over biotechnologies without any clear direction from Congress (Eisenberg 2006). Furthermore, the same year of the *Diamond* decision also marked the passage of the Bayh-Dole Act (1980), albeit under what the medical journalist Harriet Washington deemed 'morally questionable circumstances', namely, in the last hour of the last day of the last Congressional session, which suggests that the bill's passage depended crucially on the relatively low percentage of Congress present (Jacobson 2011). However, currently judges and politicians endorse a 180-degree change in legislation, disregarding potential variation in the law on patents and debate about other potential forms of genome ownership. By presenting only two alternatives – *either* the genome is a natural part of the human being and therefore unpatentable *or* the genome is raw material open to those financially equipped to prospect for it and thereby file for a patent – a third way is overlooked (for more on this dichotomy, see Fuller 2002: chap. 1).

To be sure, the patentability of the genome is endorsed by scientific and corporate establishments. They argue that patentability is a reward for years of costly research and an incentive to develop further products and treatments. Nevertheless, this argument is deeply flawed. Before 1980 when the Bayh-Dole Act allowed US universities to transfer their patents to commercial entities, genes and living organisms were already patentable (e.g. Pasteur had patented a yeast in late nineteenth-century France) and genetics was developing under the roof of universities. In the latter case, universities certainly spent vast amounts of money and time to develop the patentable materials. However, Bayh-Dole effectively decreased the amount of time and money spent on genetic research for corporations to acquire genetic patents. They could now get patents previously held by universities for a price smaller than what universities had invested to obtain the patent in the first place and then patent genes that do not require extensive scientific intervention and hence, do not attract extensive costs (Jacobson 2011). On this basis, the venerable science journalist, Daniel Greenberg (2007) has argued that universities would both serve their own interests, narrowly construed, and the public interest more broadly if they themselves more aggressively exploited the commercial potential of the Bayh-Dole Act. We take on board this shrewd point in our quest for non-standard bearers of intellectual property rights.

It is worth considering the extent to which businesses have exploited the Bayh-Dole Act. Patents for gene fragments before 1999 could have been granted for the mere fact of isolating them without specifying their utility, in line with the famous quote from *Diamond v Chakrabarty*, 'Congress intended statutory subject matter to include anything under the sun that is made by man'. Companies were filing patent forms without knowing what the product actually does in nature. Fortunately the United States Patent and Trademark Office's (USPTO) guidelines, finalized in 2001, have partially rectified this situation by requiring that the usefulness of a gene must be shown before it becomes patentable (US Department of Energy 2010). Companies see patentability as a way of capitalizing on their resources at a low cost by either laying their hands on a resource without properly testing it beforehand or carrying out their research in developing countries where regulatory checks are much less stringent than in the USA or Europe. It is a valid policy concern that corporations should not act irresponsibly, but it is regrettable that corporate behaviour has pushed courts and politicians to turn their backs on the patentability of genes.

Whether genes can or cannot be patented should depend neither on their classification as either natural or artificial, nor even on knowing the entirety of their functions. Rather it should depend on our understanding of their potential for development. We have not only obtained information about genes but also discovered that genes are information in their own right, key to our biological understanding of self. Therefore, conceptualizing how genes can be appropriated has fluctuated between considering them as a capital resource to a piece of intellectual property. Neither characterization is perfectly accurate nor completely wrong. However, given their centrality to the promotion of human evolution and social progress, genes should be subject to some sort of legal recognition comparable to registration. According to Richard Dawkins' (1976) 'selfish gene' thesis, the evolution of life is all about genes constructing organisms (including humans) as vehicles for reproducing themselves. From a genetic standpoint, we do not own the genes but the genes own us. Genes provide the structure of human existence by informing the traits on the basis of which humans expand collectively as well as cultivate and enhance their individual sense of self. In short, we enter the world as genetic debtors by virtue of possessing capacities that we did not earn. But what are the potential legal implications of such ideas?

Much turns on the legal concept of 'alienability'. While genetic material may be, at least in a weak sense, 'alienated' from the organism bearing it (as naturally happens in biological reproduction but also whenever a DNA sample is taken), the organism cannot be 'alienated' from its genetic material. You cannot give away your genes in the same way that you can give away your kidney and still remain who you are. That sense of 'inalienability' incurs obligations because both your individual existence and the future of the species are tied to a common fate that is dependent on what you do in your lifetime. As we have seen, this view has been historically associated with eugenics, often with an eye to 'conserving' the gene pool from corruption (e.g. through bad marriages and over-reproduction of undesirable classes). But in these pages we have also witnessed a more proactive approach to our genetic obligations, rendering them more a matter of investment than simple conservation.

The argument for treating the genome as a form of intellectual property amounts to the claim that individuals by default own only their bodies but not their genes. By analogy, people own their technical skills but not necessarily the means of producing those skills. People trained in a trade or a profession can sell the products of their training but not the training itself, which requires further licensing and accreditation. Similarly, genetic ownership corresponds to a duty of cultivation that must be met by the potential owner. To be clear, by 'genetic owner' we mean holder of a patent that entitles the holder to temporary excludable control over the use of the patented object, be it process or thing. Thus, someone else (including the state) could hold a patent on your genes, while you formally own your body, with the result that the patent holder may restrict your sphere of action. But this point needs to be put in perspective. The law already restricts the sphere of action available to your body when it is in the public interest (e.g. no violence to another body) as well as the conditions – if any – under which you can sell parts of your body. In the case of genetic ownership, we are calling for a re-evaluation of the terms on which intellectual property rights are granted for genetic material, potentially expanding the range of candidate 'owners' and the character and objects of their 'ownership'. And while we believe that patents may be especially effective, and focusing on them helps to simplify the current discussion, we are open to other legal instruments and mechanisms to regulate what are soon likely to become integrated 'information markets'.

The legal recognition of genes should begin by acknowledging a default right of their bearers to own and use them. As the material substratum of human autonomy, our genes provide a tangible basis for thinking of ourselves as literal works in progress. Thus, to deny that genes can be registered/patented would be to deny this special status. However, questions remain as to who should own the patent and the form of ownership that is most appropriate. For example, if the genome is considered a common property of all humans by the virtue of some common goal of collective self-reproduction, then the state should act as a protector of genome and enact provisions for individuals and companies to acquire licences for its use. Crichton (2006), a best-selling novel, ended with just such a proposal in order to pre-empt *de facto* corporate ownership of human lives that transpires in his slightly fictionalized view of the future. Alternative solutions may also be considered. However, traditional readings of one legal criterion of patentability – that the proposed invention cannot be 'obvious' or 'common' (i.e. already available to the public before the patent is filed) – fails to do the relevant work (United States Patent and Trademark Office 2011: Part II, Chap. 10). After all, latest estimates reckon that all humans share 99.5% of their genome (National Institutes of Health 2011). Yet, from this overwhelming sense of prior commonality, it neither follows that one knows the implications of that commonality for individual lives nor the respects in which and the effects to which individuals differ within the remaining 0.5% range.

In short, the natural–artificial and common–uncommon distinctions are of no help in determining whether patents can capture the spirit of genetic stewardship, which is to say, an alignment in the aims of patent law and genetic stewardship. To seek such an alignment, we need to look at the conceptual and motivational underpinnings of patent laws. The Constitution, which provides the basis for the US patent law system, states that the Congress has the power to promote the progress of science and useful arts. Q.T. Dickinson, former Under-Secretary of Commerce for Intellectual Property has argued that the US founding fathers intended a flexible, technology-neutral system from gearshifts to genomics for the advancement and spread of inventions (Dickinson 2000). For them, the purpose of assigning intellectual property was to give due credit with the aim of promoting human virtues and, more generally, humanity as a species capable of creating and determining its biological and technological environment. In the case of genetic stewardship, this would mean that humans are empowered to adjust the environment to enhance genome and acquire more knowledge about genome to put it to better use. Moreover, intellectual property is a way of developing a resource, protecting design and enhancing utility (Moore 2011). Even though agreement is unlikely to be reached on whether genes are any one of these things – a resource, design or a tool enhancing utility of other objects – the purposive, functional character that genes share with these qualities requires that we accept genes to belong in the realm of intellectual property (cf. Fuller 2002: chap. 2).

Even though the law on patentability of genes has been in a state of flux, attracting many misplaced questions, genes can be patented because their function necessitates their legal recognition. Once we accept that we have either the right or the duty to make the most of our gene pool, the system of patent law may provide the framework for a *de facto* national population policy based on the idea of genetic stewardship. In the past, eugenic policies were enacted with the explicit aim of producing the biggest yield or best quality of genes in the pool, which then mandated that, say, certain foetuses or even individuals were eliminated to create a 'quality' gene pool. In contrast, genome patenting would fall under the category of technologies geared for the non-eliminative betterment of human condition. The patentability of genes encourages both the exploration of one's heritage and the taking of control over one's fate.

However, stewardship of one's genetic heritage involves managing risky judgements about what constitutes a 'good' or 'bad' use of genetic resources. Consider the decisions taken by American courts vis-à-vis fairness in distribution of risks and benefits of genetic manipulation. As in the recent case of Myriad Genetics, they prohibited further patentability, presuming that those who seek patents would be companies that (again presumably!) would exploit society. However, we maintain that genes should be patented precisely to address this particular concern. Maybe not in the form that we see today – maybe the state should own all the patents (*à la* Crichton 2006), or maybe we should adopt a subtler licensing strategy, 'hedgenetics', as discussed below, in order to manage patents – in both cases, with an eye to attributing worth to human material while removing restrictions and precautions on its use.

Worth is closely tied to identity and provides the stimulus for concern, action and betterment. If we did not already confer value on human genes, if we did not see them as constitutive of our sense of self-identity and self-determination, it would be irrelevant who has final say over them. However, by giving genes legal recognition and legal worth we are acknowledging that protection needs to be afforded. Here is the crux of the matter: patents allow for exclusivity of use by the 'inventor' and can potentially deprive an individual from whom the genes have been extracted. However, the reverse logic should be applied: Once we establish that genes should be patented, society must decide who can own a patent and what restrictions on it can be placed. *Not allowing genes to be patented opens doors for abuse: no regulation means that no laws can be invoked when abuse occurs.* Hence, genes should be patentable to express worth of genes and to enable mechanisms for protection of values and dignity.

Lastly, the genome should be patented to maintain legal certainty. As previously mentioned genetic material and organisms have been patented in America for more than 30 years now. Since the law does not act retrospectively, the judgements surrounding the Myriad Genetics case do not invalidate already acquired patents but cast a shadow on their exploitation. If Congress does not uphold the patentability of the genome, everyone involved – scientists, corporations, citizens – will be confused as to what form of ownership of the genome, if any, is possible. The status quo is not welcome: a decision needs to be made about what follows in the event that lawmakers disallow patenting of the genome. And even once new legislation is introduced, it will take time for the general public and private enterprises to comprehend the new rules and adjust accordingly. In the meanwhile both our sense of duty of genetic stewardship and perhaps even genetic research itself may be arrested. A new legal concept is needed. We propose *hedgenetics* to encapsulate a collective right to gene ownership compatible with the duty of genetic stewardship.

The concept of genetic stewardship entails a substantive obligation to cultivate one's genome. On the surface the obligation resides in the individual who must engage in a kind of Foucauldian self-policing in order to enhance his or her genetic core. It appears reasonable from the standpoint of physical access – genes reside exclusively in an individual. To highlight this technical aspect most legal systems require a sane adult to give a legally valid consent before undergoing invasive medical procedures or participating in scientific research. In principle, no democratic government has the right to extract human genetic material or information therein without their consent, even with the best intention of cultivating and enhancing his or her genetic potential. However, in the recent years the law on consent has been relaxed, such that an increasing number of procedures are performed based on the court's mere presumption of a patient's mental state, as in the case of forced caesarean sections on pregnant women (Sheena 2005). Unsurprisingly, then, some medical lawyers have made the case that consent has been treated as a tool to enforce arbitrary decisions of judges and doctors in the name of public policy. Thus, it is not unforeseeable that the notion of consent might be further relaxed to accommodate genetic procedures in the event of a court being involved. This general trend provides grounds for enabling, if not outright encouraging, people to take active responsibility for cultivating their genome – if only to resist such incursions, as per the liability rule approach to law discussed at the start of this chapter.

Moreover, the individual has a right to privacy and a right not to be tortured and exposed to demeaning treatment (Council of Europe 1950: Articles 8 and 3, respectively). British law draws the line at 'grievous bodily harm'. Generally speaking, penal codes and criminal law, informed by human rights law, inform debates about how much pain is acceptable to inflict on a human being. This makes any positive obligation to cultivate one's genetic core restricted by norms of respect for one's physical and mental integrity. One can cultivate one's genome within those constraints. This might sound obvious until we take into account the reality of genetic modification practices. Ordinarily people cannot just walk into a laboratory and perform a genetic test on themselves. They require professional support to engage in science of any sort, including tests that the professionals themselves might regard as 'routine'. Often these genetic experts are employees of commercial enterprises who run tests for financial gain, beyond whatever service they provide to the tested individual. Understandably, then, an individual willing to express his genetic stewardship by undergoing a procedure in a typically alien scientific environment is protected by the concept of human rights and human dignity. Yet, this protection overrides any desire that he and the scientists performing the procedures might have in violating it. In short, the parties are not free to negotiate a mutual agreeable arrangement.

Thus, serious questions arise about the scope for exercising any notional duty to genetic stewardship under the current regime of human rights legislation. As a point of reference, consider the legal obstacles placed in the way of disposing of one's body in live kidney donation cases: e.g. demonstration of absence of commercial interest, consent of the donor and counselling of both parties: 'The clinician responsible for the donor must refer the matter to the Human Tissue Authority (HTA), which before making a decision must consider reports from independent qualified persons who have interviewed both the donor and the recipient and the regulations.... The donation must be approved by a panel of three members of the HTA if the donor is an adult who lacks capacity or a child or when donation is paired, pooled, or is non-directed and altruistic' (Mason and Laurie 2013: 573–4). To be sure, eliminating human rights from the equation could result in too much death or incapacitation, which would defeat the entire purpose of genetic stewardship. But judicial attitudes are still very likely to recoil at the idea of consenting to the prospect of 'grievous bodily harm verging on torture' in order to undergo an experimental procedure.

In this context, an interesting precedent is provided by British courts in their consideration of cases in which a sane adult, in pursuit of sexual pleasure, has engaged in sado-masochistic gay activities with the consent of all parties involved and yet was convicted of bodily assault (*R v Brown* 1994). This decision, albeit excessive, is a barometer of judicial, if not societal feelings, towards the willing expression of one's individuality or, as the transhumanists like to call it, 'morphological freedom' (Bostrom 2005). If judges are reluctant to allow consenting adults to engage in risky sex acts, what hope is there for them to allow consenting adults to engage in risky experiments in their home in the sort of devolved 'do-it-yourself' fashion advocated by synthetic biologists as the proactionary future of 'open source' science more generally (Church and Regis 2012)? Courts, in the first instance, protect bodily integrity, even at the cost of self-fulfilment and subjectively defined quality of life.

Seen from outside the human rights frame, the nature of individual responsibility implied in genetic stewardship is again apparent given the genetic interdependence of all living things. Crudely put, if you do not pull your weight genetically, you may become a burden to the genetic commons of the planet, which is the scope in which some anthropologists have begun to think about biodiversity governance (Oldham et al. 2013). Interdependence requires that every individual respects others' duty of genetic stewardship and fulfils his or her own duty of stewardship. Humans share a vast percentage of their genome not only with their own species but also with animals, especially mammals (National Institutes of Health 2005; Elsik et al. 2009). The genetic interdependence within our species is even clearer when we look at inheritance – every individual possesses half of their genetic make-up from both parents and has limited scope for deliberate genetic modification unless scientific procedures are involved. Thus, making the best use of one's genome means technological enhancements and scientific investments that are compatible with promoting the interests of the rest of humanity, which indirectly will benefit the entire biosphere – even if not a particular species. Reversing the line of argument recently pursued by a think-tank associated with the UK Green Party (Read 2012), we argue that to disregard our genetic interdependence – that is, not being proactive in its cultivation – might itself undermine the opportunities afforded to future generations, including the bio-cultural integrity of minority groups. Decisions, legal and scientific, about one's genetic intervention influence those within our cultural and emotional surroundings and require consultation, consent and collaboration from those who will be affected. In that respect, hedgenetics might be understood as a legal strategy to democratize eugenics.

Imagine a situation where genetic testing can be ordered by a court in the event of a couple's divorce to determine which parent is more suitable to keep custody of the child or a situation where companies can access children's medical records (including genetic information) to determine a parent's suitability for a job. In a de-regulated legal environment, one person's decision about genetic information can directly or indirectly determine the future of another person in ways that amplify the current asymmetries of power in society. Therefore, genetic stewardship introduces the need for human collaboration while reconciling it with individual duty to pursue fulfilment of one's genetic potential. The solution that respects both individual responsibility and collective interdependence can be encapsulated in 'hedgenetics', understood as a proactionary legal strategy that simultaneously addresses the scientific need to explore human genetic potential and the political requirements of genetic stewardship. This conceptual hybrid of hedge funds and genetics places genes in the realm of the economy as 'bio-capital', in which self-organizing groups invest in genes by pooling resources to fund research into certain genes in which they have a personal stake (Birch and Tyfield 2013).

Hedgenetics aims to empower communities by encouraging them to explore their common genetic potential and manage its exploitation. The legal literature already contains substantial discussions of the implications of genetic interdependence for issues relating to the entitlement and dispute (Laurie 2002, Jackson 2013: 409–41). Given that children inherit half of their parents' genes and often inhabit similar environments to their parents (especially girls, who traditionally live closer than their brothers to their parents to care for them in old age: Isaksen 2002), it is convenient for the family to observe similarities and exert pressure to get tests on those relatives who most manifestly display certain genetically-related traits (e.g. alcoholism). Already families today endorse 'conscious parenting', which involves genetics by, say, potential parents undergoing genetic tests before deciding on having a child, or a male predicates his responsibility for child care on knowing that he indeed fathered the child. But hedgenetics also fosters the formation of novel collectives with legal standing by allowing people to discover salient similarities in genetic make-up, perhaps based on having suffered a common fate, in terms of socially significant behavioural patterns, be they normal or deviant.

In effect, hedgenetics 'de-naturalizes', and hence universalizes, a socio-legal strategy that has already been applied by indigenous peoples' communities who wish to protect their biological heritage.

For example, the Maori community in New Zealand welcomes the collective patenting of their genes, provided that the patents respect and acknowledge Maoris and do not offend their culture and knowledge (Henry Hughes 2010). Acting on their guardianship interest ('kaitiakitanga'), the Maori lobbied the government of New Zealand to declare patents void if on the balance of interests they go against Maori interests. Maoris treat their native knowledge and genes as a means of cultivating their traditions and identity. They organize to cultivate their genetic interests just as much as they do the flora and fauna of their narrative habitats. They perceive themselves as stewards of a common ecosystem. In this respect, the Maoris are no different from families who carry out 'conscious parenting' – regardless of whether one ultimately deems their motives as selfish or selfless, their actions result in care for the genetic make-up of future generations. This is the spirit in which such self-organizing collectives would be the basic unit of genetic ownership in a hedgenetic legal regime. And here emphasis should be placed on the 'self-organizing' character of these collectives. In other words, people would be encouraged to seek out others with whomever they can make common cause through common bio-capital interests, resulting in multiple hedgenetic funds with overlapping membership.

On the basis of such demonstrated common genetic interest, social units would acquire 'standing' to invest in genes in the manner of a hedge fund. Like a standard hedge fund, where regulations limit which institutions are eligible to invest in a given fund, individuals meeting certain criteria of appropriate group membership could invest in specific genetic funds. This sustains a notion of 'horizontal equity', whereby people in like situations are treated alike. Moreover, it is consistent with common hedge fund practice where fund managers invest in the fund themselves to align their interests with the interest of the fund (Anson 2006: 123). Corporate bodies would not be directly eligible for hedgenetic status because of their already existing legal personality and pure commercial interest. However, a company could be involved in genetic stewardship by way of induction payments for individuals to participate in the company's own genetic research in exchange for 'access units' – a non-exclusive right for individuals to use the results of that research in the event that it makes a profit or a novel finding. Individuals with standing could also apply for commercial loans to finance the company's research and potentially be able to deduct interest from borrowing from their income tax.

Money pooled by members of a hedgenetic unit would be invested in a research institution of their selection, thereby prompting an independent investigation of the options, resulting in informed choices. No longer would genes be 'scouted' by physicians who have contact with patients to enrol them to gene banks, nor will the scouting be disease specific (Busby 2004). Institutions, on the other hand, would then need to be comprehensible and transparent to appeal to potential research subjects-investors. This would serve the cause of 'Protscience' flagged in this book's introduction by promoting research agendas that are accountable if not outright customized to its potential subjects-investors. The ensuing dialogues would involve both financial and non-financial considerations. Families who find no commercial gain in research into their genes may still want to carry on with research for expressive or more straightforwardly emotional reasons. Cultural groups may engage in endless research on every part of their ecosystem. Scientists may point to directions that the hedgenetic units have never considered. Research might engage in candidate gene studies whereby the relationship between some economic characteristics and genetic markers is examined or a genome-wide association study where 'genetic testers are individually tested for association with traits of interest' (Beauchamp et al. 2011: 58).

Institutions in receipt of hedgenetic funds would undergo periodic research progress and profitability checks. Hedgenetics assumes a mobile model where genome owners are not tied to one laboratory indefinitely but can transfer their genetic capital to more attractive research centres. Obviously problems may arise. The transfer of data and samples will involve transaction costs without guarantee of success. To avoid such transaction costs, and to ensure steady returns on their investments, genome owners and research centres have an incentive to cooperate for longer periods. However, to ensure that such cooperation occurs, the genome owners should institute periodic checks – perhaps on a biannual basis, given fluctuating markets. These checks would protect both investors and scientists from hasty decisions to shift allegiances, which in turn could disrupt long-term knowledge and profit flows.

Hedgenetics empowers citizens to face scientists as equals, their responsibility to others and the nature of their own individual lives.

It is a legal step short of collective genetic patents. We have been defending a paradoxical thesis: the best way to ensure the worth and integrity of the human being in the future is to make each individual legally responsible for cultivating their genome by providing the legal instruments for them to do so. As we have seen, much of the legal debate about the patentability of genes appeals to outdated conceptions of 'nature' and 'obviousness' as stopgaps against potential corporate exploitation. Without wishing to deny the potential for such exploitation, the solution is not to turn the clock back to some pre-exploitative time. Rather it is to empower individuals and groups to appropriate their own genetic material and cultivate it in the spirit of stewardship.

Yes, this is eugenics, but neither the classical state-authoritarian version nor today's *laissez faire* 'designer baby' fantasy that would allow anyone to enhance themselves and their offspring as they wish, provided that they can find the right doctor. Rather, hedgenetics would be a kind of 'participatory eugenics', a democratically accountable, legally binding version of eugenics written into the heart of intellectual property law and the regulation of financial transactions. Indeed, we would have 'genetic citizenship' come to be one of the competences that people acquire as they mature into full-fledged members of society, alongside knowledge of the basic workings of the law, the economy and the political system. However, like hedge funds more generally, hedgenetics will always be a risky investment because regardless of how finely grained our knowledge of the human genome becomes in terms of the likely functions of its various components in various biological contexts, the fact remains that even slight contextual differences can result in massive differences in genetic expression. Moreover, even the best set of legal instruments cannot ensure that all the relevant genetic groupings will receive the degree of research investment that they deserve. Here we see a role for the state in securing the genome as a public good against market failure.

The Proactionary Manifesto

'Humanity' is about more than the survival of the animal *Homo sapiens*. That point is already made in the word: 'humanity' literally means the quality of being human, independent of who or what may possess that quality. The ancient Greeks had various ways of deciding who might be so qualified: some wanted evidence of 'good character', others were satisfied with an ability to pay. That all members of *Homo sapiens* are eligible to be treated as humans is essentially an Abrahamic theological aspiration that over the past five centuries has been sharpened by science. This aspiration has typically included a desire to overcome the body of one's birth; hence the world-historic significance of Jesus – and not only in Christianity, which depicts his resurrection as having redeemed the idea of a humanity created 'in the image and likeness' of God, but also in the progressive secular world, where Jesus stands for the refusal to accept that one's starting position in life determines one's destiny. To be proactionary is, in the first instance, to identify with this progressive historical narrative, which in the secular West has been known mainly as 'Enlightenment' but in our own day is expressed as the drive to 'human enhancement'.

The drive has been expressed in many different ways. Although nowadays associated with a libertarian sensibility (e.g. the right to designer babies – if you can afford them), originally human enhancement was seen in more collectivist terms. That high medieval innovation to Roman law, the *universitas*, enabled the creation of self-sustaining corporate persons with ends of their own independent of the interests of the individuals who might serve them at any given moment. In the seventeenth century, Thomas Hobbes shifted this marginal legal category to centre stage as constitutive of what would become the signature modern way of organizing social life. Institutions dedicated to the pursuit of knowledge for its own sake (i.e. 'universities') were among the earliest expressions of this corporate sense of human enhancement. Later the state and the firm acquired legitimacy in a similar manner. With hindsight we can say that these were the original 'artificial intelligence' projects, albeit ones drafted in 'wetware' and 'dryware' (i.e. people and land) rather than today's software and hardware. Modern ideas of self-sacrifice, in both the scientific and the political arenas, were also born of this sensibility, which gave concrete expression to the metaphysical principle that the whole is greater than the sum of its parts. Existing sometimes in tandem and sometimes in tension was an ethic of individual self-purification through ascetic discipline, which typically involved intensive training of the mind and the body – again in ways that blurred sacred and secular concerns, say, in the medieval mastery of trade or an academic subject. Max Weber remains the past master of this entire collectivist tradition in human enhancement, for which the name 'sociology' is truly deserved.

However, the prospects for human enhancement have undergone a radical transformation since the Middle Ages. Originally you might be 'enhanced' through education, which in turn would allow you to upgrade your class or citizenship status. This sense of 'enhancement' is the realm within which the social sciences, especially sociology, tend to operate even today. Through minimal interference with the internal workings of your body, your speech, appearance and comportment might be changed in ways that enable a recognizably new and improved version of yourself to emerge. This vision of human upgrading, though still very familiar, was already being challenged in the late nineteenth century – that is, even before the vision had been fully realized on its own terms. Up to that point, most progressive thinkers had believed that the upgrades acquired in one's lifetime could also be somehow 'inherited' by one's offspring – either through spontaneous sexual transmission or a cultivated duty to teach others (aka culture). 'Lamarckism' is the name now given to this general socio-biological line of thought that informed the old vision of enhancement.

The new view, properly if impolitely called 'eugenics', does not deny the immediate non-invasive benefits of education. It proposes, however, that our long-term humanization requires direct physical intervention into default reproductive patterns, say, through dietary and sexual regimes. Extreme versions of this strategy in the twentieth century involved sterilization, forced migration, warfare and even genocide. But generally speaking, this eugenicist reorientation has become more sophisticated and self-applied, courtesy of the welfare state. Only conservative Christians and postmodernist followers of Michel Foucault continue to interrogate the violent heritage that informs today's talk of 'planned parenthood', 'antenatal screening', 'gene therapy', etc. In contrast, proactionaries welcome this domestication of control over the most fundamental features of human existence. We see it as providing regular opportunities for people to be reminded of their god-like power over life and death – and its attendant responsibilities. Moreover, the democratization of access to eugenic information and technologies marks a major advance over the more authoritarian versions of eugenics that were on offer for most of the twentieth century, not only in Nazi Germany but also in Scandinavia and, at least in aspiration, the United States and the United Kingdom.

At the same time, however, the relative absence of state regulation of today's bio-capital industries – partly a function of their dispersed and dynamic character – has raised the spectre of *de facto* control of our genetic capital by wealthy individuals who happen to be corporate market players. This prospect, dramatized in Michael Crichton's novel *Next*, should serve as a wake-up call to all proactionaries concerned with social justice. Our proposal of 'hedgenetics', which would confer intellectual property rights on those who have inherited certain common genes, is put forward as an example of the sort of creative legislation that is required to re-invent versions of self-ownership in 'Humanity 2.0', a world in which the 'person' is likely to be defined, as John Locke would have had it, 'forensically', which is to say, as an abstract locus of agency responsible for the management and development of certain bio-economic assets.

Clearly implied here is a radicalization of attitude towards the 'human'. Gone are the days when John Rawls (1971) could persuasively ground an elaborate defence of the welfare state on the intuition that anyone uncertain about their place in society would prefer to live in one that promises to protect its members from the worst outcomes. For Rawls and his still many die-hard fans, the just society is self-evidently precautionary. In marked contrast, a proactionary world would not merely tolerate risk-taking but outright encourage it, as people are provided with legal incentives to speculate with their bio-economic assets. Living riskily would amount to an entrepreneurship of the self. Of course, society will need to be equipped to absorb the consequences of such risks, many of which are bound to be negative, at least in the short term. Greater thought will have to be given to the uncomfortable topic of 'compensation' for injury, disability and even death, which can be dealt with rationally only if there is agreement on some money-like standard of exchange, as well as agreed auditing procedures for damage claims (Ripstein 2007). It follows that insurance may suddenly become an intellectually exciting area of research with enormous metaphysical consequences for what is regarded as meaningful in life.

In any case, *pace* Rawls, justice is bound to become a harder-edged concept, one forged in the understanding that when making judgements about humanity as a whole, the ends justifies the means. The ethical question remaining is to what extent we are entitled to adopt what amounts to God's point-of-view. This point has been long recognized by 'realists' in geopolitics, who take calculated decisions about allowing their own citizens to offer up their lives in war, given the benefits that are imagined to accrue to those on whose behalf they will have fought. Informing this judgement is that value is added to the world by people voluntarily identifying with something larger than themselves and thereby consenting to serve as means to that larger end. At the philosophical level, it reconciles the two major modern ethical perspectives, Kantianism and utilitarianism, the former demanding consent on the ends and the latter efficiency in the means. For the political realist, justice consists in determining the adequacy of means to ends in this sense. After all, even if there is considerable enthusiasm to go to war, there may be less costly ways of achieving the same goals. The proactionary aspiration to 'play God', as most clearly expressed in its vision of science as 'the moral equivalent of war', extends the political realist mentality from the war room to all walks of life. Thus, by replacing war with science, it may be possible to inspire people to absorb many of the same costs – including personal harm or death – in a less violent manner and for more reliable benefits to those who matter to them, now and in the future.

Nevertheless, liberal societies of the sort that Rawls justified are being challenged in a still more fundamental way. A non-conformist Christian such as John Locke could easily imagine all humans as natural equals by virtue of our having been born of the same God. In that case, we have only ourselves to blame (courtesy of Adam) for the various social barriers that we continue to impose on each other and that serve to restrict humanity's fulfilment of its divine potential. The compelling character of this rhetoric, which fuelled the American and French Revolutions, rests on a notion of common ancestry – in this case, back to God, in whose image and likeness we are created. Were we to transfer this logic to today's secular world, the result would be someone like Peter Singer (1999), for whom the normative significance of common ancestry has been naturalized, courtesy of Charles Darwin. In that case, Locke and many of his followers – not least Kant and Mill – would be cast as being among those who pose 'barriers' to the full realization of 'animal flourishing', a world in which humanity's claim to welfare extends only to our common animal capacity to experience pleasure and pain, but not to some notion of 'higher intelligence' that might lead us to turn a blind eye to animal suffering out of a self-regarding sense of species privilege.

While proactionaries instinctively stick with Locke, the opening of the legal floodgates for risk-taking will result in unprecedented strains in our notions of a 'maximally inclusive' (what Locke would have called a 'tolerant') society. The problems here go beyond the exercise of 'morphological freedom' touted by transhumanists. After all, some people may wish to forgo, as a matter of principle, any opportunity for enhancement. They may be perfectly happy with the 'Humanity 1.0' ideal enshrined in the United Nations Universal Declaration on Human Rights. Alternatively, as we have just seen, a perverse spin may be put on such an opportunity by trying to construct societies in which people and animals live as 'equals' (and not simply by implanting human features in animals). These are equally 'risky' options that also conform to the letter, if not the spirit, of the proactionary principle. And even transhumanists will be faced with their own extreme members who understand the bodies of their birth as merely convenient platforms for launching into cyberspace, the realm where they believe that their true identities can fully unfold. Thus, the proactionary challenge to classical liberalism is how to promote a climate of tolerance in a society whose members are no longer compelled by a sense of common ancestry and are inclined to veer into increasingly divergent futures.

Consider two deep histories of the human condition that might vie for followers in a proactionary world. They would agree on both the naturalist premise that we are products of evolutionary forces and the supernaturalist premise that we are destined for a life that radically breaks with that of our ancestors. Strikingly, both histories incorporate what is now normally taken as strictly 'biological' features of our humanity into political economy.

One way to flesh out this narrative – in classic eugenic fashion – is in terms of improving and extending the bodies of our birth. In the future so envisaged, an enhanced version of us might look much as we do now but, say, live much longer or have much better memories. Getting the science right, however, would only be the beginning of an unprecedented transformation in the organization and governance of the human condition. After all, until now the normative structure of society has presupposed that members of *Homo sapiens* possess a life cycle of roughly equal and finite duration. Indeed, persistent differences in life expectancy between classes continue to justify outrage about 'social inequalities'. Yet, once the assumption of a common life expectancy is suspended, then questions arise as to the conditions under which age should be rewarded or discounted in order to maintain social order. Why not have citizens vote for leaders only once in their lives, namely, at an 'age' when their personal investment in the past and the future are in 'equilibrium'? This would mean that politicians are called to account more frequently but by fewer and different people each time. Whether in each instance the relevant people would be of the same chronological age is an open question.

The other way to flesh out the proactionary narrative – in the spirit of Ray Kurzweil – is to treat our biological heritage as a prelude to our true fate, which involves merging with the technology that up to this point has enabled us to project an enduring human image on the world. In effect, *Homo sapiens* is a harbinger for '*Techno sapiens*', a creature that is destined to acquire a kind of cyborg-like existence, if not abandon the carbon substratum altogether. In this brave new world, biological evolution is the prehistory of technological evolution, in which the various organic species are understood as the (divinely?) invented products of constrained trial-and-error. These flora and fauna are exemplars to inspire human technological ingenuity to take forward in a more consolidated and focused way. The field of engineering called 'biomimetics', which explicitly treats life forms as technological prototypes, would replace ecology as the disciplinary context for understanding the relation of organisms to their environment (Benyus 1997). From this perspective, the ecologist's inclination to restrict the 'natural' functioning of an organism to the environments that historically stabilized its survival is like the tendency of some intellectual historians to limit the meaning of a concept to the contexts in which it first acquired a stable identity (e.g. Skinner 1969).

Richard Dawkins' (1976) 'selfish gene' interpretation of evolution has been widely criticized for suggesting that organisms – not least humans – are simply vehicles for the propagation of genes. From a strictly proactionary standpoint, the only real problem with this proposal is the involuntary nature of our servitude to genetic propagation. In sociological terms, Dawkins makes us out to be 'evolutionary dupes'. However, proactionaries do not have a principled objection to seeing one's body as a means for realizing a larger end, especially if it enables what one regards as an improved expression of our humanity. In this respect, proactionaries accept a quite literal understanding of genes as 'bio-capital': namely, as currency through which one sort of thing is exchanged for another, resulting in the mutual enhancement of the traders. A legal framework characterized *inter alia* by a shift to *an understanding of entitlement in terms of liability rather than property rules, a declaration of a 'right to science' and the institutionalization of hedgenetics as a legally recognized category of social activity is designed to facilitate this shift to a more proactionary self-understanding of what it means to be human.*

The transhumanist philosophy of 'ableism' offers important insight into the emerging normative horizons of the proactionary world. It argues that in a world where mental and physical enhancement will become more commonplace, we should expect people's self-understanding to shift to an existential condition of 'being always already disabled', if only by virtue of not having undergone the latest popular treatment (Wolbring 2006). As a result, something that modern society had taken for granted for so long – a stable sense of norms in terms of which people might orient their actions and on the basis of which they might claim entitlement or register grievance – would disappear. In effect, you will be forced either to become 'autonomous' in the true Kantian sense of a law unto yourself or have your identity forever buffeted by countervailing trends in how people extend themselves. Put in Existentialist terms, you will be either infused with Nietzsche's will to power or beset by Kierkegaard's *Angst*. The burden on proactionaries will be to design welfare states that tolerate such a diversity of human conditions, whereby what some judge to be an enhancement to their capacities is taken by others to be a sign of disability. Failure in this task might result in formally recognized sub-speciation: Apartheid 2.0.

An instructive precedent here is provided by the Christian Democratic versions of European welfare states that politically redeem the idea of an inherently fallen humanity (Daly 2006). The relevant theology is Calvinist, which implies that one's status in life is at best an indirect indicator of one's spiritual fate. If you are willing to remain Christian under such risky circumstances, then you are invited to treat each person as worthy of care and consideration, in particular as a lesson in living from which everyone else may learn. Hence, we might glean God's sense of justice, or 'theodicy'. The Christian character of this orientation is meant to conjure the image of Jesus as someone whose humiliation and death inspired millions in perpetuity (the closest pagan analogue is Socrates). There are, however, many secular variants that run the gamut of existential commitment, all designed to add value to a state of being that otherwise might be regarded as degraded if not worthless. For example, the blind and the deaf, precisely due to the additional constraints on their sensory modalities, have developed innovative world-views and cultures that expand our understanding of humanity's 'morphological freedom' (Fuller 2006: chap. 10). Not surprisingly, many of them have responded to technologies aimed at restoring sight and hearing or pre-empting the birth of blind and deaf people with a resistance comparable to that surrounding past 'negative eugenics' proposals.

Proactionaries do more than 'make a virtue out of necessity' in terms of the spontaneous generation of variously endowed humans. They also promote and extend such variation as a strategy of self-transcendence, which might be expected of a creature that has so far fallen short of realizing its full divine potential. Thus, in a proactionary regime, many conditions for which past welfare states were designed to prevent or mitigate as 'too risky' would now be encouraged and compensated. These would be seen as providing a rich source of data and understanding for extending our collective sense of humanity, on which individuals may draw as they choose for their own experiments in living, from which others may then draw their own lessons. Karl Popper famously held that the value of a scientific theory lies in its capacity to reveal the limits of the theorist's relationship to the world, which historically has sometimes resulted in a fundamental reorientation in our epistemic horizons. Entrepreneurs deploy new products with analogous intent at the ontological level, namely, to get people to expand their sense of self by reorienting their consumption patterns. In both cases, the failures are at least as instructive as the successes in defining, as the German idealists would have it, the boundary between what we are and what we are not. In the end, it is a concern for how that fundamental distinction is drawn that marks us as 'human'.

Legislation and Cases

Association for Molecular Pathology v. Myriad Genetics. (2013). 569 US 12–398.

Bayh-Dole Act, Pub. L. No. 96–517, codified at 35 USC §§200–12.

HARVARD/ONCOmouse [1991] EPOR 525.

HOWARD FLOREY/Relaxin [1995]. EPOR 541.

Kirin-Amgen, Inc. v Hoechst Marion Roussel Ltd. [2004] UKHL 46.

Moore v Regents of the University of California (1990). 51 Cal. 3d 120; 271.

R v Brown (1994) 1 AC 212.

re MB [1997] 2 FCR 541.

Rochdale v C [1997] 1 FCR 274.

Sidney A. Diamond, Commissioner of Patents and Trademarks, v. Ananda M. Chakrabarty, et al. (1980). 447 US 303 100 S.Ct. 2204, 65 L. Ed. 2d 144, 206 USPQ 193.

Bibliography

Agassi, J. (1975). *Science in Flux.* Dordrecht NL: Reidel.
Anson, M. (2006). *The Handbook of Alternative Assets.* Hoboken, New Jersey: John Wiley & Sons.
Armstrong, K. (2009). *The Case for God: What Religion Really Means.* London: The Bodley Head.
Arnhart, L. (1998). *Darwinian Natural Right: The Biological Ethics of Human Nature.* Albany NY: SUNY Press.
Barben, D., Fisher, E., Selin, C. and Guston, D. (2008). 'Anticipatory Governance of Nanotechnology: Foresight, Engagement and Integration'. In: E. Hackett et al. (eds), *Handbook of Science and Technology Studies.* Cambridge MA: MIT Press, pp. 979–1000.
Barrow, J. and Tipler, F. (1988). *The Anthropic Cosmological Principle.* Oxford: Oxford University Press.
Bashford, A. and Levine, P. (eds) (2010). *Oxford Handbook of the History of Eugenics.* Oxford: Oxford University Press.
Bauman, Z. (1991). *Modernity and the Holocaust.* Cambridge UK: Polity.
Beauchamp, J. et al. (2011). 'Molecular Genetics and Economy', *Journal of Economic Perspectives,* 25 (4): 57–82.
Becker, E. (1973). *The Denial of Death.* New York: Free Press.
Becker, G.S. (1964). *Human Capital.* Chicago: University of Chicago Press.
Benassi, D. (2010). 'Father of the Welfare State?' *Sociologica* 3: 1–20.
Benyus, J. (1997). *Biomimicry.* London: Penguin.
Birch, K. and Tyfield, D. (2013). 'Theorizing the Bioeconomy: Biovalue, Biocapital, Bioeconomics or ... What?' *Science, Technology & Human Values* 38: 299–327.
Bleed, P. (1986). 'The optimal design of hunting weapons: Maintainability or reliability?' *American Antiquity* 51: 737–47.
Bloom, H. (1992). *The American Religion: The Emergence of the Post-Christian Nation.* New York: Simon & Schuster.
Böhm-Bawerk, E. (1959). *Capital and Interest: History and Critique of Interest Theories.* (Orig. 1884). South Holland, IL: Libertarian Press.
Boltanski, L. and Thévenot, L. (2006). *On Justification: Economies of Worth.* (Orig. 1991). Princeton: Princeton University Press.
Bostrom, N. (2005). 'In Defense of Posthuman Dignity'. *Bioethics* 19 (3): 202–14.
Bostrom, N. and Sandberg, A. (2009). 'The Wisdom of Nature: An Evolutionary Heuristic for Human Enhancement'. In: J. Savulescu and N. Bostrom (eds), *Human Enhancement.* Oxford: Oxford University Press, pp. 375–416.
Bowler, R. (2005). 'Sentient Nature and Human Economy'. *History of the Human Sciences* 19 (1): 23–54.
Box, J.F. (1978). *R.A. Fisher: The Life of a Scientist.* London: John Wiley.

Brague, R. (2007). *The Law of God: The Philosophical History of an Idea*. Chicago: University of Chicago Press.

Brattain M (2007). 'Race, Racism, and Antiracism: UNESCO and the Politics of Presenting Science to the Postwar Public'. *American Historical Review* 112 (5): 1386–413.

Briggle, A. (2010). *A Rich Bioethics*. South Bend IN: University of Notre Dame Press.

Broberg, G. and Roll-Hansen, N. (eds) (2005). Eugenics and the Welfare State. Lansing MI: Michigan State University Press.

Brown, M.F. (2003). *Who Owns Native Culture?* Cambridge MA: Harvard University Press.

Brush, S. (1975). 'Should History of Science Be Rated X?' *Science* 183: 1164–83.

Busby, H. (2004), *Reassessing the 'gift relationship': The meaning and ethics of blood donation for genetic research in the UK*. (PhD dissertation) Nottingham UK: University of Nottingham, School of Sociology and Social Policy. Available at: http://etheses.nottingham.ac.uk/192/1/busby_thesis_final.pdf [accessed 18 September 2012].

Calabresi, G. and Melamed, D. (1972). 'Property Rules, Liability Rules, and Inalienability'. *Harvard Law Review* 85: 1089–128.

Cañizares-Esguerra, J. (2006). *Nature, Empire and Nation*. Palo Alto CA: Stanford University Press.

Cassirer, E. (1923). *Substance and Function*. (Orig. 1910). La Salle IL: Open Court Press.

Chan, S., Zee Y.-K., Jayson, G. and Harris, J. (2011). '"Risky" research and participants' interests: The ethics of phase 2C clinical trials'. *Clinical Ethics* 6: 91–6.

Cleary, T. (2011), *Gene Patents. Should New Zealand Let the Gene Genie Out of the Patent Bottle?* (LLB Dissertation) Otago NZ: University of Otago, School of Law. Available at: http://www.otago.ac.nz/law/oylr/2011/Tom%20Cleary%20-%20LLB%20Honours%20Diss%202011.pdf [accessed 18 September 2012].

Church, G. and Regis, F. (2012). *Regenesis: How Synthetic Biology Will Reinvent Nature and Ourselves*. New York: Basic Books.

Collins, R. (1999). *Macrohistory: Essays in Sociology of the Long Run*. Palo Alto CA: Stanford University Press.

Comfort, N. (2012). *The Science of Human Perfection: How Genes Became the Heart of American Medicine*. New Haven CT: Yale University Press.

Commager, H.S. (1977). *The Empire of Reason: How Europe Imagined and America Realized the Enlightenment*. Garden City NY: Doubleday.

Cook-Deegan, R. (2008). 'Gene Patents'. In: Crowley, M. (ed.) *From Birth to Death and Bench to Clinic: The Hastings Center Bioethics Briefing Book for Journalists, Policymakers, and Campaigns*. Garrison, NY: The Hastings Center, pp. 67–72.

Corbyn, Z. (2013). 'Craig Venter: "This isn't a fantasy look at the future. We are doing the future"'. *Observer* (London): 13 October.

Council of Europe. (1950). *European Convention for the Protection of Human Rights and Fundamental Freedoms, as amended by Protocols Nos. 11 and 14* (4 November)

ETS 5. Available at: http://www.unhcr.org/refworld/docid/3ae6b3b04.html [accessed 18 September 2012].
Crichton, M. (2006). *Next*. New York: HarperCollins.
Crutzen, P.J. (2002). 'Geology of mankind'. *Nature* 415: 23.
Daly, L. (2006). *God and the Welfare State*. Cambridge MA: MIT Press.
Davies, W. (2010). 'Economics and the "nonsense" of law: The case of the Chicago antitrust revolution'. *Economy and Society* 39: 64–83.
Dawkins, R. (1976). *The Selfish Gene*. Oxford: Oxford University Press.
Dawkins, R. (1986). *The Blind Watchmaker*. New York: Norton.
De Beaune, S., Coolidge, F. and Wynn, T. (eds) (2009). *Cognitive Archaeology and Human Evolution*. Cambridge UK: Cambridge University Press.
Deichmann, U. (1996). *Biologists under Hitler*. Cambridge MA: Harvard University Press.
Dennett, D. (2006). *Breaking the Spell*. London: Penguin.
Desmond, A. and Moore, J. (2009). *Darwin's Sacred Cause: Race, Slavery and the Quest for Human Origins*. London: Allen Lane.
Dickinson, Q.T. (2000). 'Statement before the Subcommittee on Courts and Intellectual Property on the Judiciary U.S. House of Representatives, U.S. Patent and Trademark Office'. (13 July) Available at: http://www.uspto.gov/web/offices/ac/ahrpa/opa/bulletin/genomicpat.pdf/ac/ahrpa/opa/bulletin/genomicpat.pdf [accessed 18 September 2012].
Dobzhansky, T. (1937). *Genetics and the Origin of Species*. New York: Columbia University Press.
Dobzhansky, T. (1967). *The Biology of Ultimate Concern*. New York: New American Library.
Dobzhansky, T. (1973). 'Nothing in biology makes sense except in light of evolution'. *The American Biology Teacher* 3: 125–9.
Duhem, P. (1969). *To Save the Appearances: An Essay on the Idea of Physical Theory from Plato to Galileo* (Orig. 1908). Chicago: University of Chicago Press.
Dummett, M. (1977). *Truth and Other Enigmas*. London: Duckworth.
Dworkin, R. (1977). *Taking Rights Seriously*. Cambridge MA: Harvard University Press.
Eisenberg, R. (2006). 'Story of *Diamond v. Chakrabarty*', In: J.C. Ginsburg and R.C. Dreyfuss (eds), *Technological Change and the Subject Matter Boundaries of the Patent System Intellectual Property Stories*. New York: Foundation Press, pp. 327–57.
Elmarsafy, Z. (2009). *The Enlightenment Qur'an: The Politics of Translation and the Construction of Islam*. Oxford: Oneworld.
Elsik, C. et al. (2009). 'The Genome Sequence of Taurine Cattle: A Window to Ruminant Biology and Evolution'. *Science* Vol. 324 (5926) (24 April): 522–8.
Elster, J. (1983). *Sour Grapes: Studies in the Subversion of Rationality*. Cambridge UK: Cambridge University Press.
Elster, J. (1998). 'Deliberation and Constitution making'. In: J. Elster (ed.), *Deliberative Democracy*. Cambridge UK: Cambridge University Press, pp. 97–122.

Engels, F. (1939). *The Dialectics of Nature*. (Orig. 1883). Introduction by J.B.S. Haldane. Moscow: Progress Publishers.

Esfiandiary, E.M. (1973). *UpWingers: A Futurist Manifesto*. New York: John Day Company.

Evans, D. (2012). *Risk Intelligence*. London: Atlantic Books.

Extropy Institute (2004). 'Extropy Institute's Vital Progress Summit Challenges President Bush's Bioethics Council Report'. (Press Release: 19 February) Available at: http://www.extropy.org/summitpress.htm [accessed 30 July 2012].

Festinger, L., Riecken, H. and Schachter, S. (1956). *When Prophecy Fails*. Minneapolis: University of Minnesota Press.

Fisher, R. (1930). *The Genetical Theory of Natural Selection*. New York: Dover.

Fodor, J. and Piattelli-Palmarini, M. (2010). *What Darwin Got Wrong*. New York: Farrar, Straus and Giroux.

Frank, S.A. (2011). 'Wright's Adaptive Landscape versus Fisher's Fundamental Theorem'. In: E. Svensson and R. Calsbeek (eds), *The Adaptive Landscape in Evolutionary Biology*. Oxford: Oxford University Press, pp. 41–57.

Freedland, J. (2012). 'Eugenics: The skeleton that rattles loudest in the left's closet'. *Guardian* (London) 12 February.

Fukuyama, F. (1992). *The End of History and the Last Man*. New York: Free Press.

Fukuyama, F. (2002). *Our Posthuman Future*. New York: Farrar, Straus and Giroux.

Fuller, R.B. (1968). *Operating Manual for Spaceship Earth*. New York: E.P. Dutton.

Fuller, S. (1988). *Social Epistemology*. Bloomington IN: Indiana University Press.

Fuller, S. (2000). *The Governance of Science*. Milton Keynes: Open University Press.

Fuller, S. (2002). *Knowledge Management Foundations*. Woburn MA: Butterworth-Heinemann.

Fuller, S. (2006). *The New Sociological Imagination*. London: Sage.

Fuller, S. (2007a). *Science vs. Religion? Intelligent Design and the Problem of Evolution*. Cambridge UK: Polity Press.

Fuller, S. (2007b). *New Frontiers in Science and Technology Studies*. Cambridge UK: Polity Press.

Fuller, S. (2008a). *Dissent over Descent: Intelligent Design's Challenge to Darwinism*. Cambridge UK: Icon.

Fuller, S. (2008b). 'The Future Is Divine: A History of Human God-Playing'. In: A. Miah (ed.), *Human Futures*. Liverpool and Chicago: University of Liverpool Press and University of Chicago Press, pp. 6–19.

Fuller, S. (2008c). 'The Normative Turn: Counterfactuals and a Philosophical Historiography of Science'. *Isis* 99: 576–84.

Fuller, S. (2009). *The Sociology of Intellectual Life*. London: Sage.

Fuller, S. (2010). *Science: The Art of Living*. Durham UK and Montreal: Acumen and McGill-Queens University Press.

Fuller, S. (2011a). *Humanity 2.0: What It Means to Be Human Past, Present and Future*. Basingstoke: Palgrave Macmillan.

Fuller, S. (2011b). 'Why Does History Matter to the Science Studies Disciplines? A Case for Giving the Past Back Its Future'. *Journal of the Philosophy of History* 5: 562–85.

Fuller, S. (2012). *Preparing for Life in Humanity 2.0*. Basingstoke: Palgrave Macmillan.

Fuller, S. (2014). 'Neuroscience, Neurohistory and History of Science: A Tale of Two Brain Images'. *Isis* 105: 100–9.

Fuller, S. and Collier, J. (2004). *Philosophy, Rhetoric and the End of Knowledge*. 2nd edn (Orig. 1993, by Fuller). Hillsdale NJ: Lawrence Erlbaum Associates.

Funkenstein, A. (1986). *Theology and the Scientific Imagination*. Princeton: Princeton University Press.

Garcia, S.M. (1996). 'The Precautionary Approach to Fisheries'. In: *FAO Technical Fisheries Papers* (No. 350). Rome: United Nations Food and Agricultural Organization. Available at: http://www.fao.org/docrep/003/W1238E/W1238E01.htm#ch1 [accessed 30 July 2012].

Gilbert, W. (1991). 'Towards a paradigm shift in biology'. *Nature* 349: 99.

Glad, J. (2011). *Jewish Eugenics*. Washington DC: Wooden Shores Publishers.

Goldberg, J. (2007). *Liberal Fascism*. Garden City NY: Doubleday.

Goodman, N. (1955). *Fact, Fiction and Forecast*. Cambridge MA: Harvard University Press.

Gordon, S. (1991). *The History and Philosophy of the Social Sciences*. London: Routledge.

Gould, S.J. (1988). *Wonderful Life*. New York: Norton.

Gould, S.J. (1999). *Rocks of Ages*. New York: Norton.

Graeber, D. (2011). *Debt: The First 5000 Years*. Brooklyn NY: Melville House.

Greenberg, D.S. (2007). *Science for Sale: The Perils, Rewards, and Delusions of Campus Capitalism*. Chicago: University of Chicago Press.

Gregory, F. (1992). 'Theologians, Science, and Theories of Truth in 19th Century.' In: M.J. Nye et al. (eds) *The Invention of Physical Science*. Dordrecht NL: Kluwer, pp. 81–96.

Grundmann, R. and Stehr, N. (2001). 'Why Is Werner Sombart Not Part of the Core of Classical Sociology?' *Journal of Classical Sociology* 1: 257–87.

Habermas, J. (2002). *The Future of Human Nature*. Cambridge UK: Polity Press.

Hannaford, I. (1996). *Race: The History of an Idea in the West*. Baltimore: Johns Hopkins University Press.

Hardin, G. (1968). 'The Tragedy of the Commons'. *Science* 162 (3859): 1243–8 (13 December).

Harris, J. (2007). *Enhancing Evolution: The ethical case for making better people*. Princeton: Princeton University Press.

Harrison, P. (2007). *The Fall of Man and the Foundations of Science*. Cambridge UK: Cambridge University Press.

Hayek, F. (1952). *The Counter-Revolution in Science*. Chicago: University of Chicago Press.

Hecht, J.M. (2003). *The End of the Soul*. New York: Columbia University Press.

Henderson, L.J. (1970). *On the Social System: Selected Writings*, (ed.) B. Barber. Chicago: University of Chicago Press.

Henry Hughes, Patent and Trade Mark Attorneys (2010). Waitangi Tribunal Report on Flora & Fauna Claim (WAI 262). Available at: http://www.henryhughes.co.nz/Site/News_Articles_Case_Notes/Articles/Latest/Waitangi_Tribunal_Report_WAI_262.aspx#_ftnref1 [accessed 18 September 2012].

Higgins, E.T. (1997). 'Beyond Pleasure and Pain'. *American Psychologist* 52 (12): 1280–300.

Hirschman, A.O. (1991). *The Rhetoric of Reaction*. Cambridge MA: Harvard University Press.

Hogben, L. (ed.) (1938). *Political Arithmetic*. London: Routledge & Kegan Paul.

Huxley, J. (1953). *Evolution in Action*. New York: Harper & Row.

Huxley, J. (1957). *New Bottles for New Wine*. London: Chatto & Windus.

Isaksen, L. (2002). 'Masculine dignity and the dirty body', *NORA – Nordic Journal of Feminist and Gender Research* 10 (3): 137–46.

Jackson, E. (2013). *Medical Law: Text, Cases, and Materials*, 3rd edn. Oxford: Oxford University Press.

Jacob, M. (1996). *Sustainable Development: A Reconstructive Critique of the United Nations Debate*. (PhD dissertation.) Göteborg, Sweden: University of Göteborg, Department of Theory of Science.

Jacobson, B. (2011). 'Corporations Are Patenting Human Genes and Tissues – Here's Why That's Terrifying' (Interview with Harriet Washington). Available at: http://www.alternet.org/story/153203/corporations_are_patenting_human_genes_and_tissues_--_here%27s_why_that%27s_terrifying?page=0%2C1 (23 November) [accessed 18 September 2012].

Kass, L. (1997). 'The Wisdom of Repugnance'. *New Republic* 216 (22), (2 June).

Kay, L. (2000). *Who Wrote the Book of Life? A History of the Genetic Code*. Palo Alto CA: Stanford University Press.

Kelly, K. (2011). *What Technology Wants*. London: Penguin Books.

Kevles, D. (2013). 'Can they patent your genes?' *New York Review of Books* (7 March).

Kawłatow, G. (2012). 'Patentowanie Ludzkich Genow'. *Diametros* 32: 77–90.

Knight, F. (1921). *Risk, Uncertainty and Profit*. Boston: Houghton Mifflin.

Koerner, L. (1999). *Linnaeus: Nature and Nation*. Cambridge MA: Harvard University Press.

Koestler, A. (1959). *The Sleepwalkers*. London: Hutchinson.

Krier, J. and Schwab, S. (1995). 'Property Rules and Liability Rules: The Cathedral in Another Light'. *New York University Law Review* 70: 440 ff.

Kuhn, T.S. (1970). *The Structure of Scientific Revolutions*. 2nd edn (Orig. 1962). Chicago: University of Chicago Press.

Kurzweil, R. (2005). *The Singularity is Near: When Humans Transcend Biology*. New York: Viking.

Latour, B. (1987). *Science in Action*. Milton Keynes: Open University Press.

Latour, B. (2004). *The Politics of Nature*. Cambridge MA: Harvard University Press.

Latour, B. (2013). *An Inquiry into the Modes of Existence: An Anthropology of the Moderns*. Cambridge MA: Harvard University Press.

Laurie, G. (2002). *Genetic Privacy: A challenge to medico-legal norms*. Cambridge UK: Cambridge University Press.

Ledford, H. (02.11.2010). 'US Government wants limits on gene patents'. *Science*. Available at: http://www.nature.com/news/2010/101102/full/news.2010.576.html (2 November) [accessed 18 September 2012].

Leonard, T. (2005). 'Eugenics and Economics in the Progressive Era'. *Journal of Economic Perspectives* 19/4: 207–24.

Lessig, L. (2001). *The Future of Ideas*. New York: Random House.

Lovelock, J. (1979). *Gaia: A New Look at Life on Earth*. Oxford: Oxford University Press.

LSE Mackinder Programme (2010). *The Hartwell Paper: A new direction for climate policy after the crash of 2009*. London: London School of Economics. Available at: http://eprints.lse.ac.uk/27939/ [accessed 30 July 2012].

Mason, J.K. and Laurie, G.T. (2013). *Mason & McCall Smith's Law and Medical Ethics* 9th edn. Oxford: Oxford University Press.

Mayr, E. (2002) 'The Biology of Race and the Concept of Equality'. *Daedalus*, Winter: 89–94.

McCloskey, D. (1975). 'The Economics of Enclosure'. In: W. Parker and E. Jones (eds), *European Peasants and their Markets*. Princeton: Princeton University Press, pp. 123–60.

Meyer, S. (2009). *Signature in the Cell*. New York: HarperCollins.

Milbank, J. (1990). *Theology and Social Theory*. Oxford: Blackwell.

Miller, M.G. (2004). *The Rise and Fall of the Cosmic Race*. Austin: University of Texas Press.

Mirowski, P. (1989). *More Heat than Light*. Cambridge UK: Cambridge University Press.

Mirowski, P. (2002). *Machine Dreams: Economics Becomes a Cyborg Science*. Cambridge UK: Cambridge University Press.

Monod, J. (1974). *Chance and Necessity*. London: Fontana.

Moore, A. (2011). 'Intellectual Property'. In: *Stanford Encyclopedia of Philosophy*. E. Zalta (ed.). Available at: http://plato.stanford.edu/archives/sum2011/entries/intellectual-property/ [accessed 18 September 2012].

More, M. (2005). 'The Proactionary Principle'. Available at: http://www.maxmore.com/proactionary.htm.

Morozov, E. (2013). *To Save Everything, Click Here*. New York: Public Affairs.

Nagel, T. (2010). *Secular Philosophy and the Religious Temperament*. Oxford: Oxford University Press.

National Institutes of Health (2005). 'Comparing the chimp and human genomes' National Human Genome Research Institute. (31 August) Available at: http://genome.wellcome.ac.uk/doc_WTD020730.html [accessed 18 September 2012].

National Institutes of Health (2011) 'Comparative Genomics'. National Human Genome Research Initiative. (13 October) Available at: http://www.genome.gov/11509542 [accessed 18 September 2012].

National Science and Technology Council (2011). *Materials Genome Initiative for Global Competitiveness*. Committee on Technology. (June) Available at: http://www.whitehouse.gov/sites/default/files/microsites/ostp/materials_genome_initiative-final.pdf [accessed 18 September 2012].

Neumann, F. (1944) *Behemoth: The Structure and Practice of National Socialism: 1933–1944*. Oxford: Oxford University Press
Newey, G. (2012). 'I have £2000, you have a kidney'. *London Review of Books*. 34 (12): 9–12.
Noble, D.F. (1997). *The Religion of Technology: The Divinity of Man and the Spirit of Invention*. London: Penguin.
Nussbaum, M. and Sen, A. (eds) (1993). *The Quality of Life*. Oxford: Clarendon Press.
O'Connor, J.R. (1973). *The Fiscal Crisis of the State*. New York: St Martin's Press.
Oldham, P., Hall, S. and Forero, O. (2013). 'Biological Diversity in the Patent System'. *PLOS ONE* 8 (11): e78737 (16 pp.)
Oreskes, N. and Conway, E. (2013). 'The Collapse of Western Civilization: A View from the Future'. *Daedalus* 142(1): 40–58.
Passmore, J. (1970). *The Perfectibility of Man*. London: Duckworth.
Phelan, J., Link, B. and Feldman, N. (2013). 'The Genomic Revolution and Beliefs about Essential Racial Differences: A Backdoor to Eugenics?' *American Sociological Review* 78.
Pichot, A. (2009). *The Pure Society*. London: Verso.
Plehwe, D. (2009). 'Introduction'. In: P. Mirowski and D. Plehwe (eds), *The Road from Mont Pèlerin*. Cambridge MA: Harvard University Press, pp. 1–42.
Polanyi, K. (1944). *The Great Transformation*. Boston: Beacon Press.
Popper, K. (1957). *The Poverty of Historicism*. London: Routledge & Kegan Paul.
Popper, K. (1972). *Objective Knowledge*. Oxford: Oxford University Press.
Proctor, R. (1988). *Racial Hygiene: Medicine under the Nazis*. Cambridge MA: Harvard University Press.
Proctor, R. (1991). *Value-Free Science?* Cambridge MA: Harvard University Press.
Rabinbach, A. (1990). *The Human Motor*. New York: Basic Books.
Rasmussen, N. (1997). 'The Mid-century Biophysics Bubble: Hiroshima and the Biological Revolution in America, Revisited'. *History of Science* 35: 245–91.
Rawls, J. (1971). *A Theory of Justice*. Cambridge MA: Harvard University Press.
Read, R. (2012). *Guardians of the Future*. Weymouth: Green House Publications. http://www.greenhousethinktank.org/files/greenhouse/home/Guardians_inside_final.pdf.
Renwick, C. (2012). *British Sociology's Lost Biological Roots: A History of Futures Past*. Basingstoke: Palgrave Macmillan.
Renwick, C. (2013). 'Completing the Circle of the Social Sciences?' *Philosophy of the Social Sciences*.
Reubi, D. (2012). 'The human capacity to reflect and decide: Bioethics and the reconfiguration of the research subject in the British biomedical sciences'. *Social Studies of Science* 42: 348–68.
Richards, R.J. (2010). 'Darwin Tried and True'. *American Scientist* (May–June).
Ripstein, A. (2007). 'As If It Never Happened'. *William & Mary Law Review* 48 (5): 1957–97.

Rothschild, E. (2001). *Economic Sentiments: Adam Smith, Condorcet and the Enlightenment.* Cambridge MA: Cambridge University Press.
Runciman, D. (2008). *Political Hypocrisy.* Princeton: Princeton University Press.
Ruse, M. (1996). *Monad to Man.* Cambridge MA: Harvard University Press.
Ruse, M. (1999). *Mystery of Mysteries: Is Evolution a Social Construction?* Cambridge MA: Harvard University Press.
Safrin, S. (2004). 'Hyperownership in a Time of Biotechnological Promise: The International Conflict to Control the Building Blocks of Life'. *American Journal of International Law.* 98: 641–85.
Sandel, M. (2007). *The Case Against Perfection.* Cambridge MA: Harvard University Press.
Sandel, M. (2012). *What Money Can't Buy.* New York: Farrar, Straus and Giroux.
Sarewitz, D. (2011). 'Science agencies must bite innovation bullet'. *Nature* 471: 137 (10 March).
Schrag, Z. (2010). *Ethical Imperialism: Institutional Review Boards and the Social Sciences, 1965–2009.* Baltimore: Johns Hopkins University Press.
Schrödinger, E. (1955). *What is Life? The Physical Aspects of the Living Cell* (Orig. 1944). Cambridge UK: Cambridge University Press.
Schumpeter, J. (1942). *Capitalism, Socialism and Democracy.* London: Allen & Unwin.
Scott, J.C. (1998). *Seeing Like a State.* New Haven: Yale University Press.
Scruton, R. (2012). *How to Think Seriously about the Planet: The Case for an Environmental Conservatism.* London: Atlantic Books.
Seville, C. (2009). *European Intellectual Property Law and Policy.* Cheltenham: Edward Elgar.
Sewell, D. (2009). *The Political Gene.* London: Picador.
Sheena, M. (2005). *Policing Pregnancy: The Law and Ethics in Obstetric Conflict,* London: Ashgate Publishing.
Silver, L. (1997). *Remaking Eden: Cloning and Beyond in a Brave New World.* New York: HarperCollins.
Singer, P. (1975). *Animal Liberation.* New York: Random House.
Singer, P. (1999). *A Darwinian Left.* London: Weidenfeld & Nicolson.
Skinner, Q. (1969). 'Meaning and understanding in the history of ideas'. *History and Theory* 8: 3–53.
Sober, E. and Lewontin, R. (1982). 'Artifact, Cause and Genic Selection'. *Philosophy of Science* 49: 157–80.
Stehr, N. (1994). *Knowledge Societies.* London: Sage
Stepan, N. (1991). *The Hour of Eugenics.* Ithaca: Cornell University Press.
Susen, S. and Turner, B.S. (eds) (2014). *Luc Boltanski.* London: Anthem Press.
Swartz, D. (2012). *Moral Minority: The Evangelical Left in an Age of Conservativism.* Philadelphia: University of Pennsylvania Press.
Taleb, N.N. (2012). *Antifragile: How to live in a world that we don't understand.* London: Allen Lane.
Teilhard de Chardin, P. (1961). *The Phenomenon of Man.* (Orig. 1955) New York: Harper & Row.

Tipler, F. (2007). *The Physics of Christianity*. New York: Doubleday.
Titmuss, R. (1970). *The Gift Relationship: From Human Blood to Social Policy*. London: Allen & Unwin.
Turner, S. (2010). 'The Conservative Disposition and the Precautionary Principle'. In: C. Abel (ed.), *The Meanings of Michael Oakeshott's Conservatism*. Exeter: Imprint Academic, pp. 204–17.
Unger, R. and West, C. (1998). *The Future of American Progressivism*. Boston: Beacon Press.
United Nations General Assembly (1987). 'Prevention and control of acquired immunodeficiency syndrome (AIDS)' (A/RES/42/8), Official Record. New York, 42nd Session, 26 October.
United Nations General Assembly (2011). 'Political Declaration on HIV and AIDS: Intensifying Our Efforts to Eliminate HIV and AIDS' (A/RES/65/277), Official Record. New York, 65th Session, 10 June.
United States Department of Energy (2010) *Genetics and Patenting*. Office of Science. (7 July) Available at: http://www.ornl.gov/sci/techresources/Human_Genome/elsi/patents.shtml [accessed 18 September 2012].
United States Patent and Trademark Office (2011). *35 U.S.C. Patent Laws: Appendix L, Manual of Patent Examining Procedure*. (1 October) Available at: http://www.uspto.gov/web/offices/pac/mpep/consolidated_laws.pdf [accessed 18 September 2012].
Van Fraassen, B. (1980). *The Scientific Image*. Oxford: Clarendon Press.
Von Schomberg, R. (2006). 'The precautionary principle and its normative challenges'. In: E. Fisher, J. Jones and R. von Schomberg (eds), *Implementing the Precautionary Principle: Perspectives and Prospects*. Cheltenham: Edward Elgar, pp. 19–42.
Von Schomberg, R. (2013). 'A vision of responsible innovation'. In: R. Owen, M. Heintz and J. Bessant (eds), *Responsible Innovation*. London: John Wiley.
Warfield, B. (1888). 'Charles Darwin's Religious Life'. *Presbyterian Review*, pp. 569–601.
Washington, H. (2011). *Deadly Monopolies: The Shocking Corporate Takeover of Life Itself – And the Consequences for Your Health and Our Medical Future*. New York: Doubleday.
Weindling, P.J. (2004). *Nazi Medicine and the Nuremberg Trials: From Medical Warcrimes to Informed Consent*. Basingstoke: Palgrave Macmillan.
Weir, T. (ed.) (2012). *Monism: Science, Philosophy, Religion, and the History of a Worldview*. Basingstoke: Palgrave Macmillan.
Werskey, G. (1978). *The Visible College: A Collective Biography of British Socialists and Scientists in the 1930s*. London: Free Association Books.
Whately, R. (1963). *Elements of Rhetoric*. (Orig. 1828). Carbondale IL: Southern Illinois University Press.
Wheatcroft, G. (2012). 'Meet the real William Beveridge'. *Guardian* (London) 7 December.
Wiener, N. (1967). *The Human Use of Human Beings*. (Orig. 1950). New York: Avon Books.
Wilsdon, J. and Mean, M. (2004). *Masters of the Universe*. London: Demos.

Wilson, E.O. (2006). *The Creation: An Appeal to Save Life on Earth*. New York: Norton.
Wolbring, G. (2006). 'Ableism and NBICS' (15 August). Available at: http://www.innovationwatch-archive.com/choiceisyours/choiceisyours.2006.08.15.htm [accessed 13 July 2013].
Wright, S. (1977). 'Panpsychism and science'. In: J. Cobb and D. Griffin (eds), *Mind in Nature: The Interface of Science and Philosophy* (Chap. 2). Washington DC: University Press of America.

Index

Printed and bound in the United States of America